Mold Spores are Everywhere!

Black Mold

YOUR HEALTH

AND

YOUR HOME

RICHARD F. PROGOVITZ

The Forager Press, LLC

⚓ The Forager Press, LLC

The Forager Press, LLC, its employees, agents, owners and affiliates, and Richard F. Progovitz, the Author, have taken great care in the preparation of this book in an effort to provide the most accurate information available relating to the subject. However, they can accept no liability for any consequences resulting from the use of or reliance upon the information contained herein, nor for any health problems, consequences or symptoms which may arise from contact with mold, mold spores or any other fungi. Any person who seeks out mold or any other potentially dangerous fungi, or puts themselves into direct or indirect contact with it for any reason, does so at their own risk. If you suspect you have a mold infestation anywhere, the Author and Publisher strongly recommend hiring a certified mold inspector or other qualified professional to evaluate the problem.

First Printing September 2003

Library of Congress Cataloging in Publication Data
Progovitz, Richard F.

Black Mold – Your Health and Your Home – 1st ed.
Includes index

ISBN – 0-9743943-9-4

Printed in the United States of America

CONTENTS

MISSION STATEMENT

This book was written to aid people in determining if the indoor air they breathe is harboring toxic mold spores and, if so, what can be done about it.

SUMMARY

Growing concerns over indoor air quality combined with media reports about several high profile mold infestations have alerted the public to the potential health risks posed by toxic Black Molds.

This book proposes to separate hyperbole from the truth regarding indoor mold toxicity issues and to demystify the abundance of molds we encounter in our daily lives.

Facts about mold and mold spores that will be detailed:

- Mold spores are everywhere. Spores can be found in the air as high as 50,000 feet above sea level, to deep underground, and everywhere in between.

- Many types of mold spores can survive extreme temperatures, from boiling water and higher, to below freezing.

- Indoor mold colonies should be properly removed, as soon as possible.

- Mold colonies found in an occupied dwelling can cause health problems in susceptible people.

- Common household chlorine bleach is NOT the answer to remove mold—it is only a surface treatment that will not prevent mold from returning. Instead, borate is a superior product in mold remediation. Borate is an ingredient in the laundry detergent "20 Mule Team® Borax."

- It is impossible to eliminate all indoor mold spores.

The bottom line about indoor mold colonies is that it is better to be safe than sorry. This book will shed light on the dangers of molds and inform the reader what to do if mold is found.

ACKNOWLEDGMENTS

Over the past four decades my interest in fungi has led me to countless others who share this interest, and I have learned so much from so many.

The people who sparked my interest in fungi were my grandparents. All four emigrated from Poland to the U.S. in the early 1900s. Among the things they brought with them from "the old country" were their knowledge of mushrooms, their "ol' wives' tales," their remedies, and their ointments. By the time I arrived on the scene, they had found picking spots for edible mushrooms including their favorite, "Podpinki." This is the eastern European mushroom nicknamed the "Honey Mushroom" or in Latin, Armillaria (a complex of species).

My grandparents taught me how, when, and where to find enough to fill bushel baskets of mushrooms. They were always taken back to the cottage, cleaned, separated, cooked, and either threaded with string and dried over a Franklin wood burning stove or frozen. That was a great time in my life, and thank God I retained the knowledge so that I can pass it down to my children.

I must also thank all the family and friends who supported me—and put up with my attitudes for the long months I spent writing this book.

A special thanks to my wife Cheryl, as a Registered Nurse, she provided me with a lot of guidance about health symptoms due to molds.

I would like to thank the hummingbirds and other wildlife at my summer lakeside camp at Guestward-Ho Campground in the high mountains of south central New York, where most of this book was written. They provided me with many hours of entertainment.

Thanks to all those people who made so much information available; instructors, lecturers, scientists, teachers, mycologists, and biologists whom I have met and learned from; and all the other professionals who presented many lectures at forays sponsored by the North American Mycological Association, the Northeast Mycological Federation, the Susquehanna Valley Mycological Society and others. Mold specialists Edward Hemway and Phillip Fry also made a big difference in the way I look at mold today; their mold knowledge is uncanny. Keep teaching!

Thanks to John Haines, Senior Scientist, NYS Biological Survey in Albany, NY, for taking the time to teach me a great deal about airborne particulates. Thanks to David Fischer—mycologist, author, editor and friend who gave me guidance on this project. Thanks to Roy Reehil for his expertise, and for working with me on this book. Thanks to Melinda Ballard for reviewing the manuscript. Thanks to teachers Laura Nelson and Michelle Stahl, and Linda Hammond, Kelly Stevens, Barbara Tripp and Michael Antonetti—they all helped make this book publishable.

Thanks to Art Bell, George Noory, Whitley Strieber, Barbara Simpson, Linda Moulten Howe, Coast to Coast Late Night Radio and Dreamland. I listened to them for many hours while writing this book. They and their hosts, guests, stories, and discussions informed and entertained me many nights until five in the morning, when they finally went off the airwaves. Happy retirement Art—I will miss your late-night voice!

PREFACE

Indoor molds are surfacing as one of the 21st century's first biological wars—in homes, apartments, schools and workplaces of North America—and that's not hyperbole. Mold can be a real health threat to many individuals.

Approximately 100 species of mold are toxic, and about 15 are known to cause negative health effects in man and animals. Mold is only one factor that can affect Indoor Air Quality (IAQ). Other problem substances such as radon, asbestos, fiberglass, carbon dioxide, lead paint, and various plant allergens can contribute to health problems, but these fall outside the scope of this book.

It is impossible to eliminate all molds and mold spores inside a dwelling—but they can be minimized and controlled to a high degree. Most indoor air pollutants can be reduced or eliminated. People susceptible to mold—especially immune-compromised individuals—may relieve some of their health symptoms by improving their IAQ.

The source and cause of excessive moisture should be found and fixed immediately—within 24 to 48 hours of contamination. Once airborne mold spores have an opportunity to land on "mold food" they begin multiplying at exponential rates. "Mold food" is any substance on which mold spores can germinate, find sustenance or reproduce.

Simply cleaning up visible mold lacks remediation efficacy without discerning why it is growing there. Removing the mold's food is only one of the steps for proper mold remediation. What is the source of moisture? If you have, had, or might have any indoor water problems, keep reading; this book will give you the information that professionals use to find, eliminate, and prevent mold.

Our living habits are making mold problems more prevalent. For instance, mold problems are becoming an increasingly bigger threat due to house-to-house cross-contamination of different species of molds. This occurs when moldy materials are removed improperly and spores are allowed to become airborne, endangering surrounding dwellings and their occupants.

Is mold, a natural organism, conducting a silent and barely visible invasion on humans, animals and their dwellings?

Can mold be both a serious health threat to humans and, at the same time, a necessity for sustaining life on earth?

According to a position statement of the American College of Occupational and Environmental Medicine, "Current scientific evidence does not support the proposition that human health has been adversely affected by inhaled mycotoxins in the home, school, or office environment."

But others disagree; for instance those whose lives have been totally uprooted by molds, such as Erin Brockovich, Melinda Ballard, Ed McMahon, and their families.

PART ONE
INTRODUCTION
TO MOLDS

HISTORY OF MOLDS

Molds predate human beings by millions of years and have been documented since the dawn of recorded human history.

Egyptian hieroglyphics defined curses on invaders of their sacred tombs and burial places and many grave-robbers and archaeologists who first entered and worked there, inhaling the musty air inside the tombs, died of suspicious circumstances, including respiratory problems. The story told by those who entered King Tut's tomb was that the confined air was thick and pungent. Could this have been mold-related?

In the Old Testament, there are several references to fungi. Mold was written about and considered the leprosy of clothes and houses.

Interestingly, the following verses about molds describe mold problems occurring over 2,000 years ago. Known even then, the seriousness of infestations and the specialized remediation techniques were performed by their highly regarded priests:

"If leprosy is suspected in a woolen or linen garment or fabric, or in a piece of leather or leather-work, and there is a greenish or reddish spot in it, it is probably leprosy, and must be taken to the priest to be examined. The priest will put it away for seven days and look at it again on the seventh day. If the spot has spread, it is a contagious leprosy, and he must burn the clothing, fabric, linen or woolen covering, or leather article, for it is contagious and must be destroyed by fire.

But if when he examines it again on the seventh day the spot has not spread, the priest shall order the suspected article to be washed, then isolated for seven more days. If after that time the spot has not changed its color, even though it has not spread, it is leprosy and shall be burned, for the article is infected through and through. But if the priest sees that the spot has faded after the washing, then he shall cut it out from the garment or leather goods or whatever it is in. However, if it then reappears, it is leprosy and he must burn it. But if after washing, if there is no further trouble, it can be put back into service after another washing."

These are the regulations concerning leprosy in a garment or anything made of skin or leather, indicating whether to pronounce it leprous or not.

Leviticus, Chapter 13:47–59

Then the Lord said to Moses and Aaron, "When you arrive in the land of Canaan which I have given you, and I place leprosy in some house there, then the owner of the house shall come and report to the priest. 'It seems to me that there may be leprosy in my house!'"

"The priest shall order the house to be emptied before he examines it, so that everything in the house will not be declared contaminated if he decides that there is leprosy there. If he finds greenish or reddish streaks in the walls of the house, which seem to be beneath the surface of the wall, he shall close up the house for seven days, and return the seventh day to look at it again. If the spots have spread in the wall, then the priest shall order the removal of the spotted section of wall, and the material must be thrown into a defiled place outside the city. Then he shall order the inside walls of the house scraped thoroughly, and the scrapings dumped in a defiled place outside the city. Other stones shall be brought to replace those that have been removed, new mortar used, and the house replastered."

"But if the spots appear again, the priest shall come again and look, and if he sees that the spots have spread, it is leprosy, and the house is defiled. Then he shall order the destruction of the house — all its stones, timbers, and mortar shall be carried out of the city to a defiled place. Anyone entering the house while it is closed shall be defiled until evening. Anyone who lies down or eats in the house shall wash his clothing."

"But if, when the priest comes again to look, the spots have not reappeared after the fresh plastering, then he will pronounce the house cleansed, and declared the leprosy gone."

Leviticus 14:33–47

The references used here are from "The Book," an edition of "The Living Bible."

To summarize, the "leprosy" on clothes and in homes was mold. The priest was the expert on cleanliness; mold was unclean and considered to be leprosy.

It was understood that mold was a unhealthy invasion of clothes and dwellings, comparable to the "leprosy" of the flesh. Mold had to be inspected for; if any was found, a remediation plan was defined and implemented to decontaminate or dispose of the mold-infested objects. There is even reference to cleaning things twice—not much has changed!

OTHER HISTORICAL EXAMPLES OF MOLD PROBLEMS

One of the earliest modern attempts to describe *Stachybotrys atra* was in 1837, by a biologist named Conda. He described the mold he found on wallpaper in a home in Prague, Czech Republic.

The first reports of potentially toxic effects from *Stachybotrys* were in the 1920s and 1930s. *Stachybotrys chartarum* was deemed responsible for a disease in horses and other farm animals—mold growing on straw and grain had been fed to the animals. And in the 1940s, in Russia, the condition Stachybotrycosis was diagnosed in humans—farmers were in continual contact with mold-infested grains.

Many houses built between the 1960s and the present, with the use of energy-conserving building methods, are recognized as mold problem houses. These houses were designed and built to be fairly airtight in order to save energy, but the indoor atmosphere becomes stagnant without air drafts. This, along with a source of moisture and mold food, will incubate spores and subsequently propagate mold growth.

From 1986 to the present, numerous accounts of the condition Stachybotrycosis were reported in horses in North America.

In 1993 the New York City Department of Health convened a panel of experts to study mold in indoor environments. Remediation "guidelines" were issued in 1994 and updated in 2000. Copies are available through the NYC Department of Health website (see the Reference section in the back of this book).

Currently, the label of "Black Mold" and its association with mycotoxins is sparking much uncertainty, hysteria, paranoia, litigation, insurance claims, news coverage, health related problems, questions and concerns. Many molds produce metabolites or toxic substances, called mycotoxins, such as ochratoxin A. The seriousness of mold infestations must be understood. Myths must be put to rest and the truths must be faced with optimism to defeat indoor mold and to diminish the number of spores in an infested dwelling to attain acceptable levels. But what are acceptable levels?

Mold has become an expensive realization in the real estate world, and it won't stop. Mold is surfacing as a tough and vicious enemy of man.

STANDARDS, GUIDELINES, AND PRACTICES

The following are the latest standards, guidelines and practices that many mold experts, convinced that "no indoor mold is good mold," feel are needed in the real estate industry:

✓ Mold-testing methods must be carefully chosen, and must include both viable and non-viable methodologies. Each test must be defined, serve a specific purpose, be consistent in its results, and be repeatable. These methods should become the basis for mold infestation diagnosis (both in health and home), remediation, and litigation, and gain acceptance as standards in the IAQ industry and governmental agencies.

✓ Mold-testing equipment, tools and kits must be standardized and defined to show their purpose. Testing times and methods—as well as calibration techniques—must be defined and standardized.

✓ Full-body protection or Personal Protection Equipment (PPE) must be standardized. From head to toe, protective clothing standards involving resistance to mold spore size, inhalation, ingestion, and clothing penetration limits must be defined.

✓ The efficiency of various delineated chemicals in killing both mold colonies and mold spores must be assessed. Chlorine bleach is not a cure-all for eliminating mold and mold spores. Fungicides and sporicides must all have documented support of their effectiveness in killing mold and mold spores, and also in the prevention of future mold infestation.

✓ Proper protocol must be established for various situations and must be generated by qualified mold experts. Strict guidelines—similar to what the military requires in their manufacturability of products by qualified suppliers in an assembly process—must be implemented. These formal plans must be as foolproof as possible, with built-in redundancies, including clearance and follow-up testing.

✓ The remediation process must align itself with proper protocol. An independent auditor in high-profile cases or serious infestations should confirm this.

✓ All mold inspectors must be certified through nationally accepted institutions. A roster of qualified instructors and their companies must be listed.

✓ "Erring on the side of caution" or "overkill" should be the accepted standard for mold-related issues.

✓ Standards and guidelines must be universally accepted, regarding construction techniques and materials with mold prevention in mind. This is especially true in flood-prone areas and in humid zones. Mold prevention should become standard practice in the new home construction industry. Landscaping, foundation, framing of treated wood, plumbing, roof, finishing carpentry, heating and cooling units, and layout should be carefully assessed if a mold-free environment is desired.

✓ Real estate transactions should include a standardized "water damage and mold" form that must be properly filled out and presented at the closing of the sale.

✓ For large mold infestations, personal testing for mold allergens should be performed in conjunction with home-inspection results to make sensible decisions on the correct protocol and subsequent remediation methodologies.

✓ Mold testing laboratories must conform to standards in their analysis and testing methods.

✓ All products that are promoted or labeled as able to kill, remove, or prevent mold or mold spores in any fashion must be tested and verified by common standards and practices, by certified laboratories or environmental agencies.

✓ Disposal and cleanup techniques and materials must be defined for the removal and transportation of mold-infested materials. However, landfills should be noted as places where mold is doing its naturally defined job—breaking down materials that can be decomposed. The transportation of infected materials from the dwelling to the landfill encompasses the real spore-spreading problem. Therefore, all mold spores, within reason, should be killed before the infected materials are removed, bagged, transported, and disposed.

Typically, industry sets the initial standards involving technologies and techniques for resolving new problems.

Liabilities, fines, responsibilities, and guidelines should be defined for all parties involved in mold-related proceedings. These guidelines must be based upon many factors, such as the cause and extent of infestation and the owners' and insurance companies' ensuing response times.

FUNGI

In 1729, Italian botanist Pier Antonio Micheli first published descriptions of fungi. This was the birth of Mycology—the study of fungi. Fungi were classified within the plant kingdom, and a new branch of botany was born.

The late 1950s and early 1960s saw Fungi reclassified within the newly established "Kingdom Fungi," the "fifth Kingdom" of living organisms. The worldwide scientific community accepted fungi as distinct organisms outside the Plant Kingdom.

Fungi are living organisms that do not require direct sunlight and do not employ photosynthesis to live, but rather require organic substances for nutrition. This is performed by the hyphae, or "roots," which secrete digestive enzymes into their food. The enzymes break down the surrounding organic material into simple molecules that are absorbed into their cells. Fungi reproduce by producing spores, which serve as the "seeds" of a fungus.

The Kingdom Fungi includes mushrooms, truffles, molds, mildew, crop rusts, rots, scabs, blights, wilts, blotches, spots, and yeasts. (Some of the names are used interchangeably.) It depends upon to whom you are talking or what you are reading or watching as to the term used. Homeowners refer to molds and mildew as the same thing. Many people use the word "mildew" in describing a particular smell, an odor, a fungal growth on garments, or a stain.

An organism can have various common names in many different languages, but it has one scientific name, which is in Latin. This Latin name is unique; it is used only for one organism. A scientific name has two parts; the genus name and the species name. Genus and species are usually *italicized* in print.

There are always new discussions and discoveries regarding the best classification of a given fungus. Some mycologists have moved downy molds, slime molds and water molds, to the Kingdom Protista, because characteristics of these types of living organisms suggest a closer relationship to amoebas than to fungi. (One such characteristic is these organisms' lack of chitin in their cell walls. Chitin is the material that forms the exoskeletons—the hard outer shells—of insects.) Other scientists propose moving water molds and downy molds to a separate "Kingdom of Stramenopila."

This is just one example of the scientific juggling that persists in the fungal world. Joining organizations such as the North American Mycological Association, the Mycological Society of America or a local mushroom club, are ways to become educated in the fungal arena.

ALL FUNGI FALL INTO ONE OR MORE OF THE FOLLOWING FOUR GROUPS:

Symbiotic
Saprophytic
Parasitic
Terrestrial

Symbiotic—When a fungus has a symbiotic association with a plant, each mutually benefits from the other—neither harms the other. Symbiotic fungi break down materials and supply the plant with mineral nutrients that it needs. In turn, the plant supplies the consistent moisture needed for the mycelium, or fungal roots, to grow, as well as other carbohydrates and chemicals that the fungus cannot manufacture.

Saprophytic—Fungal relationships in this classification receive nourishment from dead organic materials such as starches and cellulose—typically from decomposing wood, fallen branches, stumps and leaves. These are our forest floor cleaners. If not for fungi, fallen leaves would literally stack up and prevent rainwater from getting to the roots of plants, disturbing the delicate state of equilibrium in the earth's ecosystem. Many types of molds fall into this category, and some can damage or ruin food, aviation fuel, or your house. (It has been documented that more British ships were destroyed by saprophytic fungi than by enemy attacks during the American Revolution.) In the past few years, "Black Mold" has been recognized as a most notorious saprophyte.

Parasitic—Fungi that fall in this classification are destroyers of living organisms. The host material for a parasite can be anything from human tissue to a live maple tree. Some mushrooms, molds, mildews, crop rusts, rots, scabs, blights, wilts, blotches, and spots fall into this category.

A good example of a parasitic fungus is *Armillaria mellea*, also called the Honey Mushroom, Podpinki, or Oak Root Rot Fungus. When this fungus' mycelium attacks a tree, for instance, it consumes the fibrous cellulose, leaving a powdery residue.

Another example, the Parastic Club Lamb Fungus, also known as Ovine Ringworm, Lumpy Wool, or Woolrot, is a species in the genus *Trichophyton* that causes a disease in showclass sheep. These sheep are frequently washed, removing lanolin from the wool and skin. They are also frequently sheared, contributing to the disease by removing more protective wool and lanolin. *Trichophyton* is the most contagious of sheep fungus, and key to fighting the disease lies in the fact that loss of lanolin—a natural lubricant for the sheep's skin—allows the fungus to find mold food in cuts or bruises on the animal's skin.

Terrestrial – "Terra" means earth. Fungi that fall into this category grow in soil. These species, like plants, absorb nutrients from the soil. The mycelium intertwines with the moist dirt, pebbles, rocks, minerals and other materials found in soil, and sometimes produces a fruiting body such as a mushroom. Terrestrial species are generally harmless, in the context of this book.

INTERESTING FACTS ABOUT MOLD

- Mold growth begins when tiny mold spores germinate into hyphae. Hypha (pl. hyphae) is a threadlike structure of a fungus, which can grow rapidly. Many hyphae together compose the mycelium.

- Man eats food, and then digests it; mold digests food, then "eats" it. Molds create and secrete digestive enzymes, acids, and other corrosives which break down (digest) food externally, and then "eat" by absorbing food through the hyphae's cell walls. After a period of time, the remains of mold can be just dust and spores.

- Dead mold left on surfaces can release spores, and these spores can be toxic.

- Mold spores are "hitchhikers"—they cling to everything from our clothes to our pets. When we walk inside, air currents free spores, and eventually they land on a surface. If that surface is "mold food" and moisture is present, mold will begin growing.

- One square foot of mold-infested drywall can contain more than a quarter of a million mold spores.

- When surface mold is removed from or cleaned off many materials, the mold can and will return because the hyphae are not killed—many still remain viable within the substrate of the material.

- If mold lacks water or food, it becomes dormant, waits for optimal conditions, and then resumes its growth. Therefore, mold is considered "slow fire," in that it can quickly explode into a serious problem. Mold only gets worse if nothing is done to eradicate it.

- Mold does not need freestanding water to germinate and grow—moisture in the air is sufficient to sustain it. Water Activity (Aw) is the term for the amount of moisture needed for mold spores to germinate. Aw levels vary with different species of mold; a high Aw is 0.9; a low Aw is 0.1.

- There is no practical way to completely eliminate all indoor molds and mold spores. However, mold spore abundance can be drastically reduced by preventive maintenance and common sense. Eliminating all sources of excess moisture in the home will improve IAQ (Indoor Air Quality).

- An example of a beneficent mold is *Penicillium roqueforti*, which is used to make blue cheese. However, other molds in the genus *Penicillium* are not safe to eat—they don't even taste good.

- Soy sauce is made from soybeans that are fermented by an *Aspergillus* mold.

- An important ingredient in many soft drinks, especially colas, is citric acid, which can be produced by large-scale vat fermentation of *Aspergillus niger*.

- *Hormoconis resinae* is a mold that grows in aviation fuel, diesel fuel, gas, lubricants, oil, hydraulic fluids and water. The oil industry and its customers strive to prevent contamination by microorganisms in petroleum products and fuel tanks. Mold in these fluids causes organic acids, which can eat through metal, obviously undesirable in the aviation industry, as in others. A new Quantitative Colony Test, similar to Culture Plate Testing, has been developed for detecting and monitoring microbial growth contamination in oil products. The result indicates the level of contamination within a couple of days—previous testing methods gave a positive or negative result within minutes, but did nothing to quantify the degree of contamination.

- Dust mites are the only creatures known to eat some molds. Molds are not a vital part of the food chain.

THE FOLLOWING QUOTE DESCRIBES AN ANALOGY
BETWEEN A FRUIT AND ITS HOST:

"THE APPLE IS THE FRUIT
OF THE APPLE TREE,
MUSHROOMS ARE THE FRUIT
OF THE MYCELIUM,
BEER IS A FRUIT OF YEAST,
AND SPORES ARE
THE FRUIT OF MOLD!"

MOLD GROWTH REQUIREMENTS

THE FOLLOWING ARE NEEDED FOR MOLD TO GROW:

1. **Mold Spores**

2. **Water** or a source of moisture. This can be anything from a sweating water pipe to flood-like conditions.

3. **Oxygen** found in the air we breathe

4. **"Mold Food"** Wood, cotton and leather products, among others, are ideal

5. **Temperature** Mold spores can germinate when temperatures are between 32° F and 120° F; the ideal temperature range for mold growth is 70° F to 90° F.

OTHER FACTORS THAT INFLUENCE MOLD GROWTH

THE FOLLOWING CONDITIONS CAN EXPONENTIALLY INCREASE MOLD GROWTH:

1. **Relative Humidity** (RH) of approximately 50% or higher. RH indicators (analog/dial gauge or digital LED display hygrometers) can be purchased (ranging $20 to $200 or more) to monitor moisture levels. Home "weather centers" or "weather stations" usually consist of a thermometer, barometer, and a hygrometer, built into a single unit, designed to be placed on a shelf, or hung on a wall.

2. **Stagnant air** provides an added boost to mold growth by providing a stable microenvironment. Ventilation in the form of fans, open windows (depending upon the season and outside weather conditions), dehumidifiers and air conditioners work well, but only as preventive measures. Ventilated air is, in a sense, a disinfectant for mold, as it can reduce the airborne spore count.

3. **Airtight structures** specifically designed to conserve energy can adversely affect IAQ. The stagnant air precipitated by these structures provides ideal conditions for mold growth.

4. **Damp areas** left unclean and wet products (towels, rags, clothes, etc.) left indoor for two days or more, can cause mold spores to germinate and grow. (Check your children's bedrooms!)

5. Mold-infested areas that were **incompletely or improperly cleaned and remediated** can have reoccurrences in a very short period of time.

6. Mold grows quicker if there are **low nitrogen concentrations** in the air or the soil. Example: Applying higher nitrogen content fertilizer to lawns and gardens can help control excessive mold growth.

SPORES

Mold spores are so minute that they can only be viewed in detail under a microscope capable of 400X magnification or greater. Mold spores vary in size, from approximately two to forty micrometers in length, and one to twenty micrometers in width. A micrometer (μm), or "micron," is one millionth of a meter, or about 1/25,000th of an inch. (Our eyes can see only particles larger than approximately 45 μm.) Only a few molds produce spores as small as 1 μm in diameter. Spores this small can float in the air indefinitely and are very difficult to filter out of indoor air.

Mold spores are tough: they have been recovered from amber, removed, and cultured (grown) after half a million years of dormancy! Spores can survive very high temperatures—boiling water will not kill them nor will steam. Freezing only relegates spores to a state of dormancy.

Mold spores exist everywhere—even in borate and salt mines—but mold will not grow in these places.

Cobwebs are ideal locations for mold spore congregation. As spores float, air currents can guide them into a cobweb, where they become attached to a spider's sticky threads.

As mentioned earlier, mold spores are good "hitchhikers"—they are carried into our homes each time we enter. The soles of shoes are especially notorious carriers of spores.

To determine the IAQ for mold spores, a gravity test can be performed. Two to five mold spores per room, collected in a petri dish over 15 minutes, could be considered a low-level spore count. Ten or more spores may be considered a sufficiently high mold spore count to take action, especially when the outdoor spore count generates only three Colony Forming Units (CFUs). (A CFU is a hyphae-filled area that has been germinated by a single spore on a petri dish.)

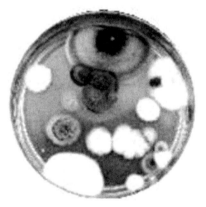

Colony Forming Units (CFUs)
of molds growing in a Petri Dish

THE SIZE OF A MOLD SPORE IN RELATIONSHIP TO OTHER TINY OBJECTS:

1. Raindrop — 600 to 10,000 μm
2. Plant spore (fern) — 10 to 80 μm
3. Mold spore — 2 to 30 μm
4. Bacteria — 3 to 5 μm
5. Tobacco smoke — 0.1 to 0.15 μm
6. Virus — 0.003 to 0.05 μm

(Key to object sizes: not to scale—diameters of circles are in microns (μm), or 1 millionth of a meter)

FIVE PHYLA OF FUNGI

*Kingdom Fungi is divided into five phyla;
each phylum includes fungi with
similar spore producing processes.*

Ascomycota—The "Sac Fungi" make up the largest of the five phyla, having approximately 35,000 known species. This phylum includes fungi such as morel mushrooms, yeasts, truffles, cup fungi, and some powdery mildews. Sexed spores (ascospores), usually in groups of four or eight, are produced within hyphae extensions in a sac-like cocoon, called an ascus. The spores are usually roundish or cylindrical in shape, with 95% of ascomycetes producing eight spores per ascus. These types of spores do not have any "tips"(apiculi) on their cell walls and are not attached to the inside of the sac. Ascospores are usually single-celled.

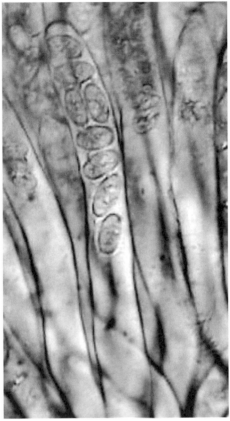

Asci sacs containing eight ascospores
from an edible Morel Mushroom
(*Morchella esculenta*) at 860X
magnification

Basidiomycota—The "Club Fungi" include approximately 25,000 known species. Their spores, called basidiospores, grow from club-shaped cells called basidia.

The "tail" or "tip" (apiculus), on these one-celled spores, is the connection point to the basidium. Spores in this class grow somewhat like cherries: multiple spores (usually four) in a group, originate from one common area, the basidia. Basidia grow on the gills of prized edible mushrooms such as chanterelles; this group also includes fungi such as boletes; puffballs; bird's-nest mushrooms; bracket, shelf and jelly fungi; corn smut; amanitas (deadly genus of mushrooms); and crop rusts which infect cereal grasses, other agricultural crops, and forest trees. Giant puffballs (mushrooms) can produce an estimated seven trillion spores per mushroom.

Chytridiomycota—These aquatic fungi include approximately 800 known species. They are called "Chytrids," and develop structures (called sporangia) that are equipped with whip-like tails for mobility. Chytrids can be saprophytes or parasites of plants and animals.

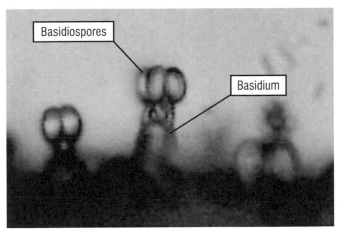

Basidia and basidiospores from an edible Blewit Mushroom
(*Clitocybe nuda*) at high magnification

Zygomycota—"Conjugation Fungi" These terrestrial fungi include approximately 600 known
species. They reproduce sexually via thick-
walled spores (zygospores) and represent species in three of the four different types of fungi
mentioned earlier— symbionts, saprophytes,
and parasites. Zygomycetes play important
roles as decomposers. They are increasingly
being used as tools for decontaminating the
environment or bioremediation; fungi are
mixed with polluted soil or water to detoxify
polluted sites.

Mitosporic or Deuteromycota—These "Imperfect Fungi" include approximately 17,000 known species. They typically reproduce asexually via spores (called conidia) on specialized hyphae, or conidiophores. Examples of fungi in this class include "good" molds from which the antibiotic penicillin was developed, and "bad" molds such as the "Black Mold"—*Stachybotrys*.

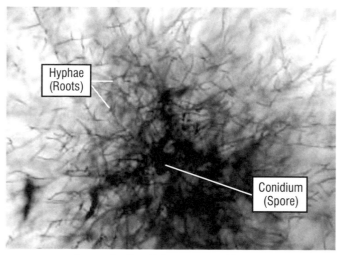

Hyphae (Roots)

Conidium (Spore)

A germinating *Epicoccum* mold spore
at 860X magnification

MOST COMMON MOLDS FOUND IN BUILDINGS

The following fungi are the most common Mitosporic indoor molds found in dwellings in North America. The molds are listed generically. In some cases, individual species are noted, but there are just too many species to list them as a matter of course.

1. *Stachybotrys*—(stack-ee-bot-ris)
    Average Spore Size: 5 x 9 μm
    Spore Shape: oval or elliptical, without a conspicuous attachment tip; smooth to gritty or warted
    Spore Color: hyaline (clear, colorless, or transparent) at first, becoming dark
    Mold Color: varies from yellow-green to brown or shiny blackish
    Sporulating Structure: hyphae produce tall (100μm to 1000μm) conidiophores (stalks) that bear clusters of ellipsoidal phialides (spore producing structures) on which slimy and sticky conidia (spores) are formed
    Common Names: "Black Mold" or "Toxic Mold" or "Toxic Black Mold"

This is one of the most dreaded types of toxic molds for a homeowner. Even if these tough molds are killed, the remaining "dead" spores still contain toxins (mycotoxins) that are dangerous. When spores become airborne, they may be inhaled with resulting health problems. These remnant spores are one of the primary reasons why proper remediation of mold infestation is crucial.

Mycotoxins can be carcinogenic, affecting the liver, kidneys, lungs, and basic cellular functions. There are approximately 200 different mycotoxins in this genus. However, the study of mycotoxins is still in its infancy, and many more species are waiting to be discovered.

Some of these toxins can contaminate grains and are heat stable (they can survive cooking). Species in this genus primarily feed on cellulose-based substances. Cotton products, basic paints, leather, and other materials will also permit the growth of *Stachybotrys*. After inoculation into an agar-growing medium, spores in this genus take 48 hours or more to germinate. Other molds grow much more quickly, often germinating within 48 hours after inoculation. The spores in this genus of molds are less frequently airborne than most other types of mold spores. *Stachybotyrs* is usually not found when performing a gravity test in a culture plate; it is usually detected by analyzing actual mold growth by using a tape lift method.

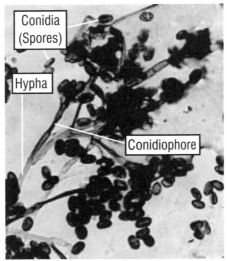

Stachybotrys mold growth at 430X magnification

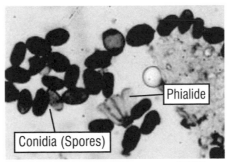

Stachybotrys mold spores at 860X magnification; the phialide shown has been separated from its conidiophore

2. *Aspergillus*—(as-per-jill-us)

Average Spore Size: 2.5–5 μm in diameter

Spore Shape: mostly round

Spore Color: the spore color in this genus varies from species to species, and from host to host

Mold Color: usually green-black, but can be yellow-green

Sporulating Structure: the entire structure resembles a toilet bowl brush; hypha grow a conidiophore (spore producing structure whose unbranched stick-shaped growth arises from a specialized foot cell of the hypha) whose vesicle (the upper swollen neck of the conidiophore) may form into a metula (a linking structure below the phialides), on where flask-shaped phialides (spore producing structures) grow; the phialides produce conidia (spores) that are attached to its tips; the conidia usually chain together and break off their phialides

Common Names: "the Green-Black Mold" or "the Toilet Bowl Brush Mold"

At least 15 known species in this genus have been discovered in homes and offices. These types of molds are commonly saprophytic in nature; that is, they feed on decaying materials such as wood, vegetables, plants, animal feed products—or most any other organic substrate.

Certain species contain carcinogenic toxins. Innumerable people will test positive with allergic reactions to this genus.

Some species cause life-threatening or generalized infections called Aspergilliosis or "Mushroom Worker's Lung Disease." Another species, *A. flavus,* is a yellow-green mold producing deadly mycotoxins; it attacks peanuts and other high-oil seeds normally stored in hot, damp areas. This species has also been reported to cause permanent lung damage called fibrosis.

Spores in this genus are so relatively light in weight that they can ride air currents without settling down for long periods of time.

Two thermal-tolerant mold species, *A. fumigatus* and *A. niger,* are pathogenic to humans. *A. niger* is a black mold commonly found on onions and garlic. It appears macroscopically (visually) as blackening on onion and garlic necks; as spots or streaks beneath the outer layers of the skins; and as black discolorations on bruised bulbs (avoid wounding bulbs during harvest, transport, and storage). Eventually, the bulb shrivels up and turns grayish-black. Store bulbs in ventilated areas, at lower temperatures (45° F–65° F.), and in a dry environment.

Interestingly, *A. niger* is commercially cultured due to its ability to produce citric acid, commonly used in the manufacture and processing of food. Citric acid can be extracted from lemon juice, but is now produced more cost effectively with the help of this mold. However, this being a very dangerous mold, many precautions must be taken to avoid inhalation of its toxic spores.

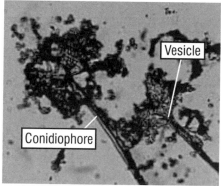

Aspergillus mold growth at
430X magnification

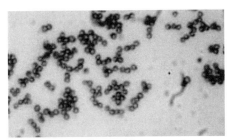

Aspergillus mold spores at
860X magnification

3. *Penicillium*—(pen-uh-sill'-ee-um)
   Average Spore Size: 2.5–9 μm in diameter
   Spore Shape: mostly round
   Spore Color: varies with its host
   Mold Color: generally bluish-green but can
   vary from white to yellow or red; *Penicillium*
   can be similar in appearance to *Cladosporium* and *Aspergillus* molds
   Sporulating Structure: resembles a paint
   brush; hyphae grow branched, short stick-
   shaped conidiophores (stalks); a metula (a
   linking structure below the phialides that
   can branch off the conidiophore on some
   species) grows; the spores emerge at the
   tips of the "bowling pin-like" shaped
   phialides; *Penicillium* means "a brush-like
   tuft of hairs," referring to its characteristic
   sporulating structure
   Common Names: "Bluish Green Mold" or
   "the Paint Brush Mold"

The spores proliferate on foods such as cheese, fruit and stored grain. Many species prefer an acidic pH for growth.

Pharmaceutical laboratories harvest a species of *Penicillium* for antibiotic use, named *P. chrysogenum*. In the name's shortened form, Penicillin, it has been a household word for many generations as a common antibiotic. Dr. Alexander Fleming discovered an antibacterial fluid excreted in the mold, in the late 1920s. Due to physician over-prescribing of this antibiotic, and improper or altered use by patients, many strains of bacteria have evolved with resistance to treatment by Penicillin. Man knows over 125 species of mold in this genus, and they were already identified by 1949.

This genus is comprised of imperfect fungi (sexual reproduction is unknown in many species). They produce vegetative spores genetically identical to the parent.

Most species in this genus can cause allergic reactions in susceptible individuals. Identifying specimens of this genus (under a microscope) to a species can be difficult, even for trained professionals.

Penicillium is commonly found in carpets and wallpaper. A high spore count inside a house can generally be traced to moist materials or decomposing human foods, which often remain unnoticed until a thorough cleaning is performed.

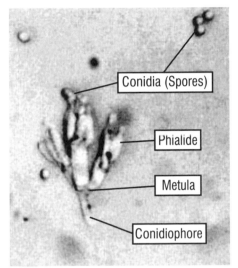

Penicillium mold growth at 430X magnification

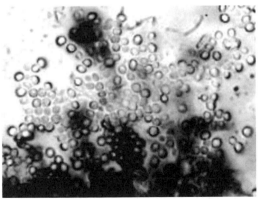

Penicillium mold spores at 860X magnification

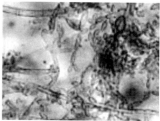

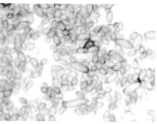

Cladosporium mold growth
at 430X magnification

Cladosporium mold spores
at 860X magnification

4. Cladosporium—(clad-oh-spore-ee-um)
Average Spore Size: 4–15 μm in diameter and up to 20 μm in length
Spore Shape: roundish when young, growing into tubular shapes, which typically have dark attachment, scars from the release areas on the hyphae
Spore Color: clear to white to cream
Mold Color: can vary from white to yellow to pink to purple to green or blackish with age
Sporulating Structure: specialized hyphae grow conidiophores (stalks) that produce conidia (spores)
Common Names: "the Dark Green Mold" Molds in this genus are both saprophytic and parasitic in nature.

Mycologists have described over 150 species. During a two-year study in 1959, biologist Kramer exposed agar plates to outside air, and found that *Cladosporium* represented about 45%

of the total spores recorded, making it the predominant genus of all airborne mold spore contaminants. *Cladosporium* is more common then *Aspergillus* or *Penicillium*.

These species are found in great numbers outdoors but also thrive indoors in shower stalls, in damp places, and in air-supply ducts.

Almost every bathtub or shower stall has, had, or will have mold growing in its corners, crevices in the grout, and on soap scum—between cleanings. Many household cleaners, including the effective 20 Mule Team® Borax and water solution (one cup of Borax to one gallon of distilled water) can kill and control *Cladosporium* mold growth in moist areas.

A ventilation fan used in the bathroom—preferably near the shower/bathtub—dries out the area quickly and reduces the potential for mold growth.

The species named *C. herbarum* has been found in timber, logs, and wood pulp products.

Another species, *C. resinae*, lives on creosote and petroleum products, including the petroleum jelly used to grease seals on pressure cookers.

Cladosporium fulvum attacks the leaves of tomato plants; colonies of this fungus appear brown to violet in color.

Furthermore, fungi in this genus are associated with asthma-related health problems.

5. *Alternaria*—(all-tur-nair'-ee-uh)

Average Spore Size: 3–10 μm in diameter and up to 50 μm in length

Spore Shape: can resemble a "caveman's club" or "a bowling pin"; conidia (spores) are large and have both transverse and longitudinal septations (cross walls or sections)

Spore Color: tan to brownish

Mold Color: from gray to brown to black

Sporulating Structure: specialized hyphae grow conidiophores (spore producing structures) that produce septate (cross walls or sectioned) conidia (spores) that are often chained together

Common Names: "Club Mold" or "Bowling Pin Mold" or "Ten Pin Mold"

Molds in this genus are very common in our daily lives. Often, mold spores are found in household dust. *Alternaria* can also be found on horizontal surfaces of building materials, such as window frames; in red mulch used in landscaping; in carpets; on textiles; on seeds; in/on unsalted butter; rotting fruits; straw; and leaves. Species in this genus can produce tenuazonic and other toxic metabolites.

These spores are some of the largest of all the molds, thus, they settle to the ground relatively quickly. The "falling rate" of a 10-μm-diameter mold spore is approximately 1 foot in 10 seconds in still air.

Species in this genus may be connected to "Baker's Asthma," which afflicts some bakery workers.

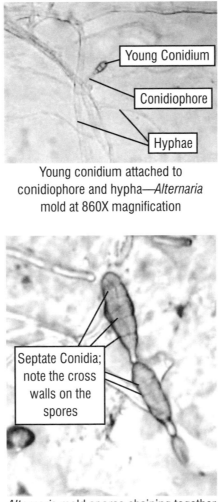

Young conidium attached to conidiophore and hypha—*Alternaria* mold at 860X magnification

Alternaria mold spores chaining together at 860X magnification

6. *Epicoccum*—(epp-ee-cock'-um)
Average Spore Size: 15–30 μm
Spore Shape: young conidia are nonseptate (do not have cross walls or not sectioned), smooth and pear shaped; becoming multiseptate (multiple cross walls or sections) and round when mature
Spore Color: tan to brown
Mold Color: varies with its host
Sporulating Structure: specialized hypha produce clusters of short conidiophores (spore producing structures) which sprout canidia (spores)
Common Name: "Soccer Ball Mold"

A slow growing, multicolored mold whose host determines the color of the mold. In an agar grain culture, the colonies range from bright yellow to pinkish orange. On most agar media, species of *Epicoccum* grow slowly and appear as white mycelium. On wood, other species in this genus can be found in colonies as small black dots. The species *E. oryzae* has been reported to infest rice. Some *Epicoccum* can excrete a yellowish fluid.

The primary role of this saprophyte is to enrich the soil through decomposition of wood, plant stems, and leaves.

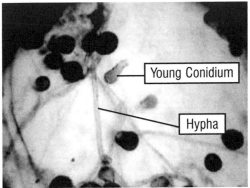

Epicoccum mold growth at 860X magnification

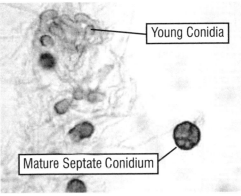

Epicoccum mold spores at 860X magnification

7. *Fusarium*—(few-sarh–ee-um)
Average Spore Size: 2–5 μm wide and 5–20 μm long
Spore Shape: normally crescent-, canoe- or sickle-shaped (macroconidia), but they can also be slightly pear- or lemon-shaped (microconidia); multiseptate (has cross walls or sections) conidia (spores)
Spore Color: hyaline (clear, colorless or transparent) to whitish
Mold Color: white to yellow to pink to purple
Sporulating Structure: specialized septate (having cross walls or sections) hypha produce single or small clusters of conidiophores (spore producing structures), which produce conidia (spores)
Common Names: "Crescent Moon Mold" or "Canoe Mold" or "Sickle Mold"

These fungi grow primarily on grains, first producing fast-growing, white, cottony mycelia, which then turn various colors.

There are species within this genus that are highly toxic. *Fusarium. sporothrichiodes* and *F. poae* produce potent trichothecene toxins. This genus of mold has bore responsibility for the deaths of approximately 30,000 people during World War II. A province in the Soviet Union was supplied with mold-infected bread made from poorly winterized grain. In the Ukraine, it has been dubbed the "Staggering Sickness" due to its symptomatic indications, which are flu-like in nature with nasal bleeding and vertigo.

This fungus has been of interest to militaries since the 1980s, when the United States accused the Soviet Union of implementing a new biochemical warfare methodology.

Lowering humidity levels and increasing ventilation of an area can control *Fusarium* molds. They are often found inside humidifiers; in some skin and nail infections; a relatively frequent agent of mycotic eye infections; and can cause other health problems.

Identification of *Fusarium* (to a species) is very difficult due to its morphological structure (variance) in spore shapes—i.e. macroconidia or microconidia.

This mold is less common than the previous six molds described.

OTHER INTERESTING FUNGI

- *Memnoniella*—(mem-non-ee-el-la) This genus is similar to *Stachybotrys* in that it grows on cellulose-based hosts. Some species in this genus can also be found on decaying leaves, from which spores readily become airborne and can subsequently enter a building.

- *Mucor*—(mhew'core) Species of *Mucor* go by the common names of "Mucor Rot," "Black Pin Mold," or "Black Bread Mold"

The toxic spores of *Mucor pusillus* thrive in temperatures between 65 and 130° F and up.

The mycelia appear like the seed-producing stalks of onions, with the spores amassed at the top, in a ball-like form. The mold is gray at first, turning blackish over time.

This species successfully propagates in piles of compost, because it is a true thermophile: an organism that lives at elevated temperatures.

Molds in this genus can also grow in old straw, grains, horse dung, plant debris, textiles, and wood chips. When wood chips are used for landscaping, heavy infestations of *Mucor* can occur and spores can easily "hitchhike" into the home when doors and windows are opened.

If infected straw is used for bedding or hay fed to horses and cattle, they can become sick.

In rare cases, *Mucor* has been known to cause brain damage.

Controlling *Mucor* mold growth is accomplished by the use of good hygienic practices (wash hands often when working with compost or wood chips), air filtration, and air circulation. It is wiser to use rocks, stones or brick chips for landscaping—rather than wood-chip mulches.

The most common species are *M. racemosus* and *M. plumbeus.*

- ***Dactylium***—(dack-till-ee-um) Species in this genus are commonly called "Soft Mildew." Dactylium is actually a mold fungus that attacks mushrooms of the genera—*Agaricus,* the common white grocery-store button mushrooms; *Amanita; Russula;* and *Lactarius.*

This rapid growing parasite at first appears as colonies of whitish, cobweb-like filaments. The mold derives its common name from the type of infestation it forms, which looks fuzzy, or patchy, on its host.

Two of its enemies are salt and the high-alkaline substance baking soda. Its spores can be killed at around a 125° F temperature sustained for about thirty minutes. The hyphae resemble a tree with branches coming off the trunk, opposite each other. The spores are produced on the tips of the smaller offshoot branches. The mold can vary in color from white to pink.

- *Candida*—(can-dee-da) This yeast commonly infects women inside the vagina. Yet, anyone can contract an internal yeast infection, called "Thrush," in any area of the digestive track. An internal infection of this yeast can be encouraged by misuse of antibiotics or by taking steroids. Though these medications are medically beneficial, misuse may affect the efficiency of the immune system, allowing *Candida* to proliferate in the human body.

ENEMIES OF MOLD AND MOLD SPORES

Mold is one of man's oldest enemies, yet we have coped with it for countless centuries. Today mold has taken on a new face: it can threaten our daily lives due to living habits, construction techniques, and other factors.

Though mold has survived for millions of years, it does have its weaknesses. If not for the balance of "Mother Nature," mold relationships would be totally out of control. The sun, wind, rain and other contingencies all play integral parts in the life of mold and mold spores.

THE FOLLOWING ITEMS HAVE A NEGATIVE INFLUENCE ON MOLDS AND MOLD SPORES:

1. **Borate**—One of the best natural enemies of mold. Isn't it compelling that one of our oldest laundry detergents (20 Mule Team® Borax) is also one of the best products for killing mold and mold spores without poisonous side-effects to humans? Simply, borate is a highly effective fungicide. A typical borate solution—one cup of "20 Mule Team® Borax" to one gallon of distilled water—can

be used on many mold-infested items without damaging them. (Avoid using tap water, which contains contaminants, bacteria, and other particulates. If necessary, filter out large, undissolved borate particles with cheesecloth, if the solution is to be applied with a spray bottle.) Distilled water is preferable, because it is pure, aiding the dissolved chemicals into porous surfaces.

2. **Fungicides** that list such ingredients as Benzalkonium Chloride or chemicals called "quats" (quaternary ammoniums) are extremely effective killers of mold. Other ingredients to look for are Di-n-alkyl dimethyl ammonium chloride or N-alkyl dimethyl benzyl ammonium chloride. Read the labels on the products carefully; do not purchase something that will not work effectively.

3. **Antimicrobial chemicals** that list metallic oxides or borates as ingredients work great for mold prevention. Some products are similar to white paint that can be applied directly on dry lumber. They are good mold killers, and are also great for the prevention of termites and car-

penter ants. Furthermore, metallic oxides are strong fire retardants and are
usually effective for at least ten years.

4. **Salt** can kill mold. For disposal purposes,
some laboratories pour salt over mold
grown in culture plates; seal them with
tape; and dispose the petri dishes. Salt
is not used to remove mold from walls,
nor as a disinfectant. A strong concentra-
tion is necessary to kill mold; diluted salt
water cannot provide an effective con-
centration.

5. **UV light** kills mold and mold spores.
This can come in the form of direct sunlight or special high-wattage UV bulbs
strategically placed in an infected area.
Hospitals use this technique to clean
contaminated rooms. Placing mold-infected items in the sun for a few hours
can kill some mold and mold spores, but
not as effectively; direct sun must shine
on a spore for a long period of time.
There are products marketed which are
installed in-line within forced-air heat-
ing/cooling system plenums. They kill
mold spores when floating by a section
housing the UV lights. These units re-

quire a substantial amount of power wattage to be effective, and can cost hundreds of dollars or more, depending upon the size of the system required.

6. **Tobacco smoke**—This is not an endorsement to smoke in your house, but homeowners who smoke tobacco products inside generally have a smaller number of mold spores in collected air samples. Smoking indoors reduces the IAQ, but smoke film left on household objects appears to be toxic to mold and mold spores. (Ironically, tobacco smoke can have both good and bad effects on man, depending upon how you look at it!)

7. **Dry indoor conditions** can keep mold spores from germinating. The RH should be controlled to be 50% or less.

8. **Improved air circulation** in houses, especially basements and crawlspaces, is key to reducing and/or eliminating molds from your home.

9. **Higher nitrogen-count fertilizers** are good for controlling mold growth on garlic and onions.

10. **Other biocides** that kill molds are isopropyl alcohol, hydrogen peroxide (3%), baking soda, and vinegar.

11. **Mothballs** emit vapors that inhibit mold growth. Mothballs have long been used to protect garments and a host of other wool- and cotton-based items. They are also excellent for use in campers and recreational vehicles during seasonal storage. Spreading mothballs or crystals on carpets, inside drawers, cabinets and storage areas will make it much more pleasant when camping season starts.

12. Common household **chlorine bleach** can kill mold and mold spores, but, in the commonly used dilution, its usefulness is diminished to the point of ineffectiveness. Using full-strength chlorine bleach is an unacceptable cleanup method, according to the Occupational Safety and Health Administration (OSHA); chlorine bleach in its concentrated form can be hazardous to your health, especially without excellent ventilation. It is a disinfectant suited to smooth surfaces but not cellulose-based products—it will not penetrate these porous surfaces.

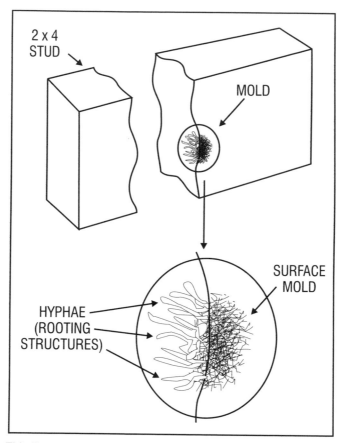

This illustrates how a mold's hyphae penetrate wood. When the hyphae are not killed, the mold returns. Household chlorine bleach cannot penetrate deep into wood; therefore, the hyphae —and the mold—will survive. A borate-based product is an effective mold killer.

PART TWO
MOLD HEALTH
ISSUES

HEALTH SYMPTOMS
DUE TO MOLDS

To best appreciate the relationship between molds and public health, one must first have an understanding of the human immune system and allergies.

In the United States, more than 50 million people—one out of five, or twenty percent of the population—suffer from asthma or allergies. According to the Asthma & Allergy Foundation of America (AAFA), allergies are the sixth leading cause of chronic disease in the United States, costing the health care system $18 billion annually. The AAFA states that hay fever has increased substantially over the past fifteen years with about 16.7 million yearly visits to health care providers, due to allergic rhinitis.

There are over 200 known indoor pollutants in the average house, with some being the source of the increase in allergies. (In the last few years, indoor pollutants have been found to be two to five times higher—sometimes 100 times higher— than outdoors.) The Environmental Protection Agency (EPA) ranks indoor pollution as one of the country's top environmental health risks.

An allergy is a hypersensitive reaction to a substance, harmless to most people but is perceived by some individual's systems as possibly dangerous. Those substances are called allergens. Certain foods such as peanuts, perfumes, soaps, chemicals, etc., can affect a person sensitive to them. Allergens, or triggers; cause the body to defend itself by producing antibodies. It is the antibodies defense/reactions that cause an allergy's irritating symptoms. An antibody is a disease-fighting protein called immunoglobin E or IgE. Everybody has a specific IgE for each allergen. For example, one will be produced to react against mold, another to react against ragweed.

The first time an allergy prone individual is exposed to an allergen, the body produces large amounts of antibodies for that specific allergen. The IgE molecules stick to the surfaces of the abundant mast cells (cells that line the respiratory and gastrointestinal tract and skin) or basophiles (circulating white blood cells). The IgE molecules circulate in the blood or remain with certain cells where they specifically counteract the invader. The next encounter with the same allergen, usually leads to an even greater reaction as the antibodies signal the mast cells and basophiles to flood the area with histamines and other chemicals. Histamines are body chemicals that can act as irritants. As histamine prepares for an attack, it inflames the surrounding tissues (nasal passages, sinuses, and eyelids), and the individual experiences typical allergy symptoms such as sneezing, itchy eyes, and runny nose.

Individuals may have family histories of allergies, but even if this is not the case, they may react to certain environmental triggers. Individuals from families in which both parents have allergies have a fifty percent chance of inheriting them. If only one parent has allergies, an individual has a thirty percent chance of having allergies. Hay fever and asthma are the most commonly inherited allergies.

One can react to many triggers indoors or out. Common indoor triggers are pet dander, feathers, dust mites, cockroach droppings, cleaning chemicals, aerosol sprays, latex, metals, tobacco smoke, milk, wheat, eggs, shellfish, chocolate, food additives, medications, and molds.

Common outdoor triggers include pollen, weeds, insect bites, insecticides, paint fumes, exhaust, pollution, smoke, cold air, exercise, and molds. With mold spores everywhere in the environment, they are impossible to avoid, and are one of the causes of chronic allergic rhinitis (along with dust mites and animal dander).

Anyone can develop an allergy anytime in life. Allergies can develop suddenly, or evolve over a period of years. Triggers can be year-round or seasonal.

When someone with allergic tendencies encounters an allergen, the body reacts to it as if fighting a bacteria or virus. The immune system's job is to protect the body from harmful invaders that can harm it by destroying or neutralizing them.

In today's world, bodies are having a tough time keeping healthy, and it makes it harder when an individual is going through physical or emotional stress. Day-to-day lifestyles, including what is put into the body, keep the antibodies working 24/7. Allergens may attack your body by land (touch), sea (swallowed), or air (inhaled). Immune systems are very vigilant in their reactions to search, identify, and destroy.

If a person is not allergic to theses triggers the mucous in the nasal passages simply moves the alien particles into the throat so that they can either be swallowed or coughed and spat out. Children are more susceptible to airborne pollutants because their lungs are still developing, and because they breathe at a faster rate than adults do.

There are many symptoms that a person can experience as part of an allergy attack, including: a runny nose; repetitive sneezing; coughing; nasal congestion and inflammation; red, itchy eyes, ears, nose and throat; ear plugging; conjunctivitis; watery eyes; nasal stuffiness; post-nasal drip; sinus headaches; sinusitis; ear infections; hives; rashes; itchy skin; swelling of lips, tongue or face; swelling (at the site of an insect bite); tightening of the throat; nausea; vomiting; wheezing; diarrhea; fatigue; "allergic eyes" or "allergic shiners" (dark areas under the eyes from increased blood flow and pooling); breathing difficulties; and, in some cases, anaphylaxis.

Anaphylaxis is a severe allergic reaction in a person previously sensitized to an allergen that may be local or systemic.

Local reactions appear at the site of exposure, and include symptoms such as skin warm to the touch, hives, swelling, and reddening of the skin. Systemic anaphylaxis can involve the respiratory, cardiovascular, and gastrointestinal systems, producing such symptoms as flushing; wheezing; difficulty breathing; increased mucous production; nausea; vomiting; abdominal cramps; increased pulse rate; a sudden drop in blood pressure; and feelings of generalized anxiety.

Systemic anaphylaxis may be mild—or it may be severe enough to cause shock when life-threatening vasodilatation occurs. Without proper medical treatment, death may occur.

Having allergies makes one more prone to sinusitis (inflammation of the sinuses), sinus infections, and asthma. Sinusitis can last just a few weeks or may become a chronic problem; it affects over 37 million Americans per year. People prone to sinus infections miss approximately four days of work per year due to their sinus problems. Bacteria, fungi, and viruses can cause sinus infections. People with weakened immune systems, such as AIDS patients, asthma sufferers, and people with cystic fibrosis are also more susceptible.

The Mayo Clinic in Minnesota has identified mold as the leading cause of chronic sinus infections. One of the most common symptoms is severe sinus headache. Antibiotics are often prescribed to treat sinusitis but may not be effective because they target bacteria, not fungi. Allergic fungal sinusitis is chronic nasal obstruction, with symptoms that include runny nose and postnasal discharge, caused by allergies to soil-based fungi (such as *Curvularia, Bisporella,* or *Alternaria*). Fungal sinusitis can also afflict the eyes and the brain.

Allergies can lead to asthma, especially in children. Asthma is characterized by hypersensitivity to an allergen that causes the tracheobronchial tree to react and cause narrowing, inflammation, and excess mucus production in the airways. This makes it difficult to breathe. Asthma can also be caused by exercise (exercise-induced asthma). Most asthma patients' symptoms include shortness of breath, tightness in the chest, coughing, and wheezing. An asthma attack can come on quickly, and it may range from mild to life threatening.

According to the EPA, in the United States, 15 million people have asthma and it affects nearly 1 in 13 school age children. Asthma is the leading cause of school absenteeism with 10 million days missed each year.

Mold spores are ubiquitous, circulated by air currents throughout our environment. When dealing with allergies, most doctors advise their patients to eliminate their exposure to the allergens. That, alas, is one request we cannot heed with respect to mold.

Unlike pollen and weed particulates, mold spores do not dissipate after the first hard frost. People who are allergic to mold may have symptoms all year long, especially in warmer, humid climates. Spore counts may vary between night and day. After a significant snowfall, outdoor mold spore counts are lower, but that does not kill the spores. Even though rain can wash some larger spores out of the air, it can also propel smaller spores into the air, which is why some mold-related allergy symptoms seem to increase during rainy days. Outdoor mold spores are more abundant in the early spring, late summer, and early fall, but can be plentiful year-round indoors. After the spring melt, molds flourish on vegetation that was killed by the winter's cold temperatures. Molds thrive in compost pits, on rotting logs, and on fallen leaves, especially in moist, shady areas. Some molds grow on grains such as wheat, oats, barley, and corn, and in grain storage areas, which are havens for molds. Indoor molds remain there year-round, even in colder areas.

Mold can be found growing in every house, but the extent of the infestation varies. Newly constructed houses are another problem; they are being built with "tighter" construction methods which reduce outside-to-inside air exchange, reduce circulation of the inside air, and cause higher indoor moisture levels. Older buildings—without the storm windows, extra insulation, etc.—are more "open" to the outside air, and have more natural ventilation. Once mold starts growing and releases its microscopic spores in a modern, energy-efficient house, the spores will become airborne and circulate through the house via the ventilation system, until they land on a moist substrate and begin growing, starting the cycle all over again.

Although mold is only one of many possible causes of an individual's symptoms related to poor Indoor Air Quality, it should be considered a potential hazard to occupants wherever evidence suggests excess moisture is present. Because it isn't possible to know what amount of mold or mold spores is safe for a particular individual, it would be wise to assume that any visible indoor mold may cause problems and should therefore be removed.

The home is not the only place where a person can come in contact with mold. The work place or schools are other locations where exposure to mold is possible, especially if one is experiencing symptoms away from the home environment. If a school or workplace has a significant group of people that are afflicted with some of the symptoms described here, you may have a "Sick Building Syndrome." Some workers at places such as farms, greenhouses, breweries, bakeries, and certain mills, as well as forestry workers and other people, whose jobs involve working with wood products or fabrics, frequently work in moldy surroundings. If a person or a family is having physically manifested symptoms from exposure to mold, they should promptly consult their physician.

People who are sensitive to molds should be aware that fungi are used in the process of making certain cheeses. Mushrooms, dried fruits, and foods containing either yeast or soy sauce sometimes elicit allergic reactions.

In 2002, a fungus based meat substitute called Quorn was approved for marketing in the U.S. The Center for Science in the Public Interest, a Washington, D.C. based health organization, called for the removal of Quorn from the market because illnesses occurred in ten percent of the people who ate it. Some Quorn labels say the product is "mushroom in origin," but that is misleading: it is made from a mold, *Fusarium venenatum*. Some of the reported symptoms from ingestion are nausea, diarrhea, hives, shortness of breath, and fainting spells. Quorn is sold in various forms including: patties, nuggets, and cutlets, and in prepared entrées such as lasagna, and fettuccine alfredo.

In the U.S. at the time this book was nearing publication, there were no EPA, CDC, state, or federal limits, standards or guidelines with respect to airborne mold spores or mold infestations. This scarcity of guidelines is primarily due to a lack of agreement between agencies regarding what constitutes a mold health problem, and also simply, because there are so many unknown variables.

According to the American Lung Association and the EPA, "Most Americans spend 90% of their time indoors, often at home. Therefore, breathing clean indoor air can have an important impact on health. People who are inside a great deal of the time may be at greater risk of developing health problems or having problems made by indoor air pollutants." The EPA also states "mold exposure can irritate the eyes, skin, nose, throat and lungs of both mold-allergic and non-allergic people."

Allergic individuals, asthmatics, young children, the elderly, immune-compromised individuals, cancer patients undergoing chemotherapy, people taking corticosteroids, and pregnant women are often the most sensitive to mold problems. However, some people not included in these groups just react more quickly to contaminants than others.

Molds have been around long before humans. One of the earliest accounts of biological and chemical weapons was in the 7th century B.C.; it is believed that the Assyrians used a fungus called *ergot* to poison water supplies. "St. Anthony's Fire" is a disease also known as ergotism; it is caused by *ergot*: *Claviceps purpurea*. This fungus most commonly attacks rye, but it can also infect other grains. When wheat fields were abandoned during the Middle Ages, rye (a "weed" grain) would often take over the fields. Rye was first used as a food product around 600 A.D. The name of the disease is derived in reference to burning sensations in the arms and legs of its victims, and the Vienna hospital that first succeeded in treating those victims was dedicated to St. Anthony. Around the year 900 A.D., when written records were first kept, severe epidemics were attributed to ergotism, every five to ten years. These epidemics were common in France at that time because rye was a crop of the poor, and the cool, wet climate provided good growing conditions for *ergot*. The rye and the *ergot* were ground together during the milling process, so the flour was thoroughly contaminated. In 994 A.D., 40,000 people in southern France died from ergotism. In 1927, 10,000 cases were reported in Russia.

Ironically, some medicines are extracted from *ergot;* Ergotamine is used for migraine headaches and Ergonovine is used to control postpartum bleeding after childbirth. *Ergot* is also the species from which lysergic acid diethyl amide (LSD) was first isolated; it was first used for treating psychiatric patients.

Certain molds, primarily *Chaetomium* and *Arthrinium*, but, to a lesser degree, *Pithomyces, Stemphylium, Torula* and *Ulocladium*, can grow in the same environment as *Stachybotrys*. If they are detected, then the home is also conducive to *Stachybotrys* mold!

Documented reports have concluded that cattle have died from hemorrhages after eating feed contaminated with *Stachybotrys chartarum*.

The "Black Molds," *Memnoniella* and *Stachybotrys*, are known to produce mycotoxins. Mycotoxins, as defined by the EPA, "are fungal metabolites that have toxic effects ranging from short term irritation to immunosuppression and cancer." There are over 200 mycotoxins known now, but this field of study is relatively young, so many more will likely be found. Mycotoxins are produced by the mold during its digestive process but are also present in spores. Mycotoxins provide the molds with a competitive advantage over other organisms. When the mold source is disturbed, whether by humans, air currents, or pets, toxic spores become airborne and can easily be inhaled. Molds and other fungi also produce volatile organic compounds called microbial VOCs or MVOCs, many of which produce distinct odors. Many professionals believe that MVOCs precipitate toxic effects in people and animals upon inhalation and skin contact.

In 1994, the U.S. Centers for Disease Control (CDC) agreed that infants in Cleveland, Ohio who inhaled *Stachybotrys* or "Toxic Black Mold" spores lived in highly contaminated, water-damaged buildings, with very high air and surface sampling levels of *Stachybotrys chartarum*. It caused bleeding in the infants' lungs—a condition known as pulmonary hemosiderosis. It is an uncommon condition found mostly in young children, but it can also occur in adults.

Since then, "Black Mold" has become one of the fastest growing health concerns in the United States. However, in 2001, the CDC changed its position. The CDC presently neither confirms nor denies the existence of a specific problem regarding exposure to *Stachybotrys chartarum* or other molds—but it does advise further research. The EPA and the Cleveland Department of Health still believe that *Stachybotrys* was responsible for the occurrence of hemosiderosis in the infants. To date, 9 of the 36 infants from these homes have died.

Other medical research has proven that inhaling or ingesting toxic mold spores can cause infections, allergic reactions, cancer, and even death. A home infected with this mold produces a potential deadly atmosphere for anyone. In May 1997, the Journal of the American Medical Association published an article called "Floods Carry Potential for Toxic Mold Disease" (Marwick, 1997). The mycotoxin—satratoxin—produced by *Stachybotrys* is poisonous if inhaled. *Stachybotrys* is a slow-growing mold that does not compete well against other more rapidly growing molds. Rarely found outdoors, it is usually hidden, and the spores are in a gelatinous mass and spread more when it dries.

Some symptoms from exposure to *Stachybotrys* include chronic fatigue; fever; chronic headaches; irritation to the eyes; inflammation and burning of the mucous membranes in the mouth, nose and throat; sneezing; skin rashes; tightness in the chest; difficulty breathing; dizziness; chronic cough; coughing up blood; cold- and flu-like symptoms; hair loss; memory loss; nausea; vomiting; diarrhea; bleeding in the lungs; nosebleeds; asthma; and death.

Cladosporium is common in both indoor and outdoor air samples. Sensitive individuals exposed to high levels of this mold are more likely to develop allergy or asthma-like symptoms.

Cladosporium werneckii causes the condition called Tinea nigra. Tinea is a fungal disease occurring on the body; it is commonly called "ringworm." This fungus will live on dead tissues of hair, nails, and skin. The chances of this fungus growing on a body increases in the presence of poor hygiene, continually moist skin, and minor injuries. Tinea nigra is a fungus infection that affects the skin of the palm and/or sole. Tinea corporis is ringworm of the body. Tinea capitis is ringworm of the scalp. Tinea pedis is a fungus infection of the foot, which is commonly referred to as "athlete's foot"—caused by fungi that thrive in warm, moist areas. The usual symptoms of athlete's foot are itching, stinging, burning, and cracking and peeling of skin on the foot and between the toes. Athlete's foot is only mildly contagious, but it is frequently spread by foot-to-floor contact in public showers and at swimming areas, or by sharing towels or bath mats. Ringworm can be contracted from direct contact with other people, animals (many pets have ringworm), or from the soil. It can usually be remedied by proper use of an over-the-counter anti-fungal agent, but a physician should be consulted and may prescribe an oral anti-fungal medication.

Some of the approximately 200 known *Aspergillus* species produce toxins called aflatoxins, citrinin, gliotoxin, patulin, sterigmatocystin and ochratoxin. Aflatoxins are one of the more potent carcinogens in the environment. Both aflatoxin and ochratoxin have been classified by the National Toxicology Program as carcinogenic to humans. The maximum allowable level of total aflatoxin in food commodities is set by the FDA at 20 parts per billion. The maximum level for milk products is even lower, at 0.5 parts per billion. This toxin is poisonous to humans if ingested and toxic to the liver. Ochratoxin damages the kidneys and liver, and impairs the immune system. Citrinin is a nephrotoxin, linked with renal damage, vasodilatation and bronchial constriction. Gliotoxin is an immunosuppressive toxin. Patulin causes hemorrhaging in the brain and lungs and has been linked to grape and apple spoilage.

There is also a medical condition called allergic bronchopulmonary aspergilliosis, which causes the patient to wheeze, have a low-grade fever, and cough up brown sputum and mucus plugs. It can invade the lungs, the brain, the kidneys, and heart valves.

Aspergillus spores are frequently found in hospitals, and are a big concern for bone marrow transplant patients. They have been found to cause corneal (eye), otomycotic (ear), and nasoorbital (nose) infections. In collecting samples of *Aspergillus*, whether cultures, tape lifts, or air samples are taken, the spores are almost indistinguishable from *Pencillium* spores on cultured samples, unless conidiophores are present.

Fusarium is a genus of common molds that produce Trichothecene toxins, which target the skin and the circulatory, alimentary, and nervous systems. Also, it is found in some skin, eye and nail infections. If infected grain is ingested, symptoms may include nausea, vomiting, diarrhea, dermatitis, and extensive internal bleeding.

Candida infections are fungal infections caused by yeasts. They include infections of the skin, nails, vagina, mouth, and throat (thrush). These fungi can also invade the bloodstream in immunosuppressed, burn, and dialysis patients.

Candida is present in the body's normal flora, but when there is a disruption in the body, the likelihood of infection increases. *Candida* can multiply if a person is pregnant, obese, or diabetic, if the person has cancer, immuno-suppression, or is taking either antibiotics or high doses of corticosteroids.

Candidiasis of the nails can occur in individuals who frequently have their hands in water or wear nonporous gloves. People who are overweight or incontinent of urine are at risk because it can grow in constantly wet skin folds. Personal hygiene can play an enormous role in the prevention of these types of mold infections.

HOW MOLDS ENTER OUR BODIES

Mold spores can invade or enter human and animal bodies in five ways. Molds can grow internally in such places as the nose, sinuses, brain, lungs, eyes, ears, toes, liver, and intestines.

1. **Inhalation**—As we breathe, airborne mold spores enter our lungs with virtually every breath. There is no practical way to eliminate this exposure. There are fewer mold spores in the outdoor air during the winter, but there can be a greater number of airborne spores indoors at that time.

2. **Skin contact**—Touching mold-infested sources with fingers or any other part of the body can bring about health problems in an individual. The biggest danger presents itself when an individual comes in contact with mold and then inadvertently touches his face and either inhales or swallows spores. Practicing good hygiene, and keeping hands away from your face can avert this danger.

After bathing, dry all parts of the body thoroughly, especially crevices or creases and between toes.

3. **Ingestion**—Many types of molds are literally eaten in our everyday consumption of food in meals, snacks, and beverages. Molds grow on cheese, breads, pastries, meats, fresh spices, vegetables, fruits, and nuts. The presence of visible mold on food is evidence that the mold has been growing for some time. At first, the growth is so limited that it cannot be seen with the naked eye, or even with a magnifying glass. Visible colonies appear on food almost overnight.

4. **Eyes**—Mold spores can be wiped into our eyes by our hands and clothing. Spores can also settle on our eyes during such daily routines as house cleaning.

5. **Ears**—Spores can enter the body through the ears in the same way they enter via the eyes, and mold can grow inside the inner ears if spores penetrate the tympanic membrane. Again, good hygiene can help prevent mold growth.

High Profile
Mold Infestation Cases

Three publicly documented, high profile cases provide compelling evidence that water problems can lead to mold growth, and that mold and mold spores can cause health problems. All three involve the "Sick Building Syndrome."

Case #1: Erin Brockovich

Erin Brockovich, the environmental activist whose $333 million lawsuit against Pacific Gas & Electric Co. for leaking toxic chromium-6 into the groundwater of a small California city and who was the subject of a major film starring Julia Roberts, has a mold-infested mansion. Employing the royalties from the movie, she purchased a million-dollar home that became infested with mold. She has filed a lawsuit naming the previous owner and the builder, charging that each played a role that caused water damage. Testing revealed the dreaded *Stachybotrys* mold. Brockovich has spent $250,000 remediating the mold infestations, and is now stuck with a lemon of a home. She can't sell the house without revealing the status of the building, and who would want to buy it? She became a mold activist and is currently pushing for California legislation that may set the first mold standards in the country.

CASE #2: MELINDA BALLARD

A professional couple purchased their "Shangri-La"—a 72-acre estate west of Austin, Texas. It turned out to be one of the worst nightmares of their lives.

Melinda, her husband Ron, and their three year-old son Reese developed many short- and long-term health problems because water leaks were not fixed, and the building materials were not dried out within 48 hours of the leaks' first occurrences.

The house's copper water lines developed a series of leaks in 1998. By December, the hardwood floors in several rooms began to warp. Three months later, in March of 1999, the entire family, along with two house support members, began suffering from fatigue, headaches, dizziness, and respiratory and sinus problems. In short time, their conditions worsened to the point where they were coughing up blood.

Initially, the house was not suspected as being the source of their health problems, and everyone tested negative for allergies. By chance, on an airplane flight, Melinda's coughing led the person sitting next to her to ask if anything was wrong. The man, an Indoor Air Quality consultant, then asked if there had been any recent water damage to their house. Putting two and two together, he made arrangements to visit the house. He took samples of the visible mold, which was subsequently identified as *Stachybotrys*. The residents were then advised to evacuate immediately. With only their clothes on their backs, they moved into a hotel. Their house was quarantined. Not only was their health in serious jeopardy, but also their home and all of their belongings were effectively stripped away from them.

Melinda filed a lawsuit against her insurance company for failing to remedy the water and floor problems despite repeated warnings from a contractor. A jury awarded Ballard $32 million in their first-party bad-faith lawsuit. The jury agreed that the insurance company mishandled the family's claim for water damage and the ensuing toxic mold exposure. The house will have to be de-contaminated, razed and rebuilt. The family's health is improving, but some effects could be permanent—only time will tell.

CASE #3: ED MCMAHON

Ed McMahon, Johnny Carson's longtime on-air sidekick, claims that his insurance company botched a simple repair on a pipe that burst and flooded the den of his Beverly Hills mansion. In April 2002, he filed a lawsuit for $20 million. He charges that the contractor painted over visible mold, and that they failed to provide him with environmental reports related to the mold infestation. McMahon claims that he, his wife, and their housekeeper were sickened, and his dog died as a result of the mold. During the remediation process, it was also discovered that the air-conditioning ducts in the contaminated area were not sealed off, and the mold spores were spread throughout the house as a result. This case settled out of court in April 2003.

Part Three
Molds and
the Home

HOW MOLDS
ENTER OUR HOMES

Mold spores can invade and enter our homes in a variety of ways. Because spores are very lightweight, air currents provide a super-highway for them to travel endlessly and aimlessly. Some mold spores are so lightweight that they may never settle down to earth! Other spores are heavier and will eventually settle on something.

When a dog walks through tall plants, spores practically "jump" to attach themselves to him because of simple static electricity. An animal's paws and fur are very good places for which spores to cling. The dog comes in the house, lies down on the carpet, and falls asleep—releasing thousands of spores as he runs inside, wags his tail, and rolls on the carpet. The air circulates, and, very soon, new spores are everywhere inside the house. Every time the dog goes out and comes back in, more spores hitch a furry ride into the house.

But simply eliminating pets does not mean eliminating spores. Every time a house door is opened or closed, more spores come in, and of course some go out. Many of the spores in the house settle and don't become airborne again until they are disturbed (for example, when the homeowner is dusting or vacuuming).

Every time people return from vacation or a trip—whether to the nearby grocery or to a distant part of the world—they bring home spores. We are, in short, "stocking" our homes with the spores of many species of molds, and they are patiently waiting for the chance to germinate, thrive—and cause problems! Most of those spores never do germinate—but many of them do.

The possibilities of how spores can enter one's life are endless. Man has even taken spores beyond the earth's atmosphere in his spacecraft, both intentionally (for scientific studies), and, of course, unintentionally. Every human sheds many pounds of skin each year— estimates are as high as 30 pounds per year for an average-size adult. In a case that attracted worldwide media attention, spores hitched a ride to the Mir Space Lab to find a veritable feast of shed human skin cells; after a while, the space lab started to wreak from mold growing on the cosmonauts' shed skin!

SOURCES OF INDOOR MOISTURE

Moisture in the home can facilitate mold infestations, if not promptly remedied (within 24 to 48 hours after an event). Every homeowner should be aware of potential sources of mold-promoting indoor moisture.

1. **Floods**—Any room, especially cellars or basements damaged by flooding, are prime candidates for mold growth. Rising and settling waters transport mold spores that occur naturally in the soil during floods. Houses in flood zones are particularly vulnerable to mold invasions.

2. **Pipes**—Water from leaky pipes can drip or trickle onto many different types of mold food promoting spore germination. This moisture could lead to expensive repairs, unless the leak is stopped immediately. Inspecting for mold growth and getting the area cleaned and dried quickly can help limit growth. The quicker the problem is fixed, the better. Simple condensation on pipes can drip

on mold food and create a problem as well. Wrapping water pipes with insulation will prevent pipes from sweating.

3. **Roofs**—Roof leaks are particularly hard to pinpoint, and expensive to repair. Roof leaks usually do not indicate precisely where the leak actually originates. Water travels by gravity but can find its way horizontally along boards and other building materials before reaching the spot at which a leak becomes evident. Roofs cannot always be inspected from the inside, so finding the leak can be difficult. Whenever or wherever roof damage is evident, prompt repair is vital!

4. **Faucets**—Bathrooms, kitchens, outdoor faucets, laundry rooms, spa and sauna rooms all present the potential for water leaks and repeated spills. These areas must be kept dry and clear of any sign of visible mold growth.

5. **Toilets**—Another case of Toxic Black Mold infestation in a residential home in upstate N.Y. was caused by a leaky toilet seal, which allowed water to seep through the floor and walls in a large area of the house. The family endured

allergic reactions and asthma symptoms for many months. When they finally received help and identified the problem, they moved out. After a few weeks, their conditions started to improve. At the time of this writing, the family was still living in a hotel while getting some of the repairs done, simultaneously trying to get their insurance company to pay the total repair costs—which will run about $100,000 to fix what had been a $250,000 house. This family has been totally uprooted by mold! A leaky toilet has ruined their good health, property, and belongings.

6. **Refrigerators**—Condensation inside refrigerators can be found on food; it promotes mold growth on vegetables and other foods. Refrigerators that have built-in water-and-ice machines have waterline connections in the back. Moving refrigerators around for cleaning can put stress on the fittings, leading to drips that may go unnoticed for long periods of time. The water can drip on the floor and mold can quickly develop, especially if there is a significant dust build-up under the appliance. When refrigerators with waterlines are moved, the floor

should be checked for any leaks several days after the unit has been placed back in its position. Also it must be ensured that the waterline has enough play in it, and that it has been properly connected at both ends. The waterline should be fastened somewhere on the back of the refrigerator to strain-relieve the pipe connection. The installation directions that come with the refrigerator will indicate proper procedure or a professional can help.

Mold also can grow on refrigerator doors' sealing gaskets; inspecting and cleaning these gaskets periodically is a necessity.

7. **Dishwashers and Washing Machines—** These appliances are notorious for water leaks; dishwashers also release steam through vents in the front. Water-supply connections and wastewater hoses should be well maintained.

In dishwashers, water also leaks through the seals between the door and the front of the unit. With proper installation and upkeep, these appliances can be problem-free; otherwise, slow leaks in inaccessible spots can go unnoticed for years.

8. **Water Heaters**—Water contains many different chemicals—minerals and believe it or not, corrosive elements, which not only shorten the life and efficiency of a unit but which can also start a leak, typically at the bottom of the tank. An efficient water heater generally lasts less than ten years. It is far better to be on a preventative maintenance schedule with water heaters than to wait for them to leak unnoticed.

Inspecting the water connections and the bottom of the tank periodically can limit leakage. Areas that have "hard water" are especially prone to heavy settlement of particulates in their water tanks. Be careful draining a few gallons of water from the drainage valve, located on the bottom side of the unit. A surprising volume of calcium, and other solids that have settled to the bottom of the tank will come out with the water—when drained on a monthly or bimonthly basis. The drainage valve must not leak after closing it. Periodic observation of the water connections is good preventive maintenance. Proper installation is vital: due to the heater's typical out-of-the-way location, leaks often go unnoticed for weeks or even months.

9. **Sewers**—All homes have sewer pipes for the disposal of wastewater from sinks, bathtubs, toilets, and other appliances. Sewer pipes in most homes drain directly into municipal pipes and from there, down to the community's sewage treatment plant. Some rural homes drain sewage into underground holding tanks and/or leaching systems. Any sewer can back up under conditions such as flooding and clogs. When this occurs, there is usually a foul odor—potentially toxic fumes are present. Besides being an immediate health risk, this is a mold warning sign—mold growth will occur in the very near future if the area is not promptly cleaned, dried, and sanitized.

10. **Basements, Cellars and Crawlspaces**— These areas are notorious havens for mold growth. Water from a number of sources can produce severe, chronic infestations.

   Proper drainage outside of the foundation can help prevent water seepage into a basement. Cracks in the concrete walls and floors are a common route for water entry. Stone or cinder-block foundations should be coated with concrete or another suitable material to help keep water out. Heating and cooling air conditioning units should never be installed in crawlspaces, where there is not much air circulation. Carpets should never be used on concrete floors; moisture will form naturally under carpets, and, when there is other water infiltration from flooding or spills, the carpet will often never dry out completely.

Cellars are often used like attics—for storing an accumulation of belongings. If boxes are stacked up against the walls covering the concrete floor, and water enters the area, many of the things stored in those cardboard boxes and other containers will quickly become mold food. Investing in watertight storage bins will help protect both your home and those old family possessions.

11. **Fruit and Root Cellars**—These are great places to store foods for long periods of time, but moisture must be controlled and air ventilation is vital. These precautions especially hold true if onions, garlic, potatoes, fruit, and other foods are stored for extended periods. Many different kinds of mold can infest a harvest. Someone once said, "One bad apple in the basket will spoil the whole bunch."

Perhaps a farmer, who realized the dangers of mold in crop storage, originated this phrase.

12. **Moisture Vapor**—The following are significant sources of moisture vapors, and average quantities released (for six of those sources) inside the house:
    A. Breathing—one quart/day/person
    B. Showers—one half pint/shower
    C. Baths—one tenth pint/bath
    D. Mopping floors—1½ pints/10 square feet
    E. Dishwashing—1 pint/load
    F. Cooking—one to three quarts/day
    G. Humidifiers
    H. Houseplants
    I. Fish aquariums
    J. Pets
    K. Exposed soil in crawl spaces

Potential sources of indoor moisture vapor are numerous—they are not limited to merely those cited above.

ITEMS THAT CAN BECOME "MOLD FOOD"

Over time, mold can grow on almost anything if moisture is present, but the potential is exacerbated if there is high mold nutritional value in the substrate. Although mold spores are everywhere, mold will not grow on some things if they are clean: rocks, brick, metal, glass, or concrete. If mold is found on any of these, the mold is most surely growing on dirt, dust, or on paint.

Don't be fooled into seeing mold growth where none exists: A buildup of oil or soot on walls and ceilings from candles or from kerosene or oil heaters can appear very similar to mold.

Two species of *Stachybotrys*— *S. atra* and *S. chartarum*—require either cellulose or cotton based substrates. Cellulose and cotton are used in the manufacture of an endless variety of products, including drywall, gypsum board, sheet rock, ceiling tiles, furniture, newspapers, books, magazines, paper bags, cardboard boxes, wall paneling, flooring, carpeting, draperies, insulation, dynamite, and, of course, any kind of wood.

Firewood should not be stored inside the house—the wood often contains mold spores. Do not use wood chips or bark chips as mulch around the house—they can enhance the appearance of a property, but they are usually loaded with mold! In the past several decades, more and more people have begun using wood chips for landscaping—yet another reason why mold spores are becoming increasingly abundant inside homes.

Cotton-based items that can become mold food include clothes, blankets, sheets, pillowcases, sleeping bags, rags, towels, rugs, furniture, hats, potholders, and footwear, to name just a few. All cotton-based items in the home should be kept dry!

Other favorite mold foods include dirt and grime on shower stalls and bathtubs. Flooring beneath bath mats can be a problem area for mold growth if the area is not cleaned and dried frequently. If a non-slip mat is used on the floor of a tub, it should be removed and dried, after each shower/bath.

MOLD CAN GROW ON MANY THINGS; THE FOLLOWING ARE SOME COMMONPLACE ITEMS THAT MAKE SUPERB MOLD FOOD:

Food—Old food left in the back of the refrigerator for a week or more often yields a moldy surprise. Anything from meat to potatoes can grow mold. Eventually, that opened jar of spaghetti sauce will show a fuzzy, greenish-colored mold. Always examine leftover food before reheating and/or eating it.

People do not usually get sickened from eating moldy food, because most molds are minimally toxic and because the taste is often unpleasant—molds are not part of a healthy human diet. Allergic reactions pose the greatest danger here. Nuts, rice, grain, meat, tomatoes, cauliflower, other vegetables—indeed most foods—should always be inspected for mold contamination prior to consumption.

Purchase only those fresh food items that you will use in the near future—do not buy a bag of nuts, put them in the cupboard, and forget about them for months. Also, do not just take for granted that stores only sell fresh, uncontaminated food—buyer, beware!

Juices and Soda—By leaving an inch or so of soft drink in a bottle with the cap on it and setting it aside for a few weeks, one can see the effects of mold. In as little as one week, mold will appear inside the bottle. While the cap was off, spores found their way in—perhaps even directly from an individual's own mouth. Spores are everywhere.

Bird Food—Bird food, left in a bag on a concrete floor, will eventually grow mold. Most birds do not like mold either; the seeds should be inspected for mold contamination prior to restocking bird feeders. Mold can be prevented from infesting a bag of "bird seed" by it being used entirely within a reasonable amount of time—a few weeks—and by storing it in a dry place.

Grains—Historically, dangerous mold and mold spores have contaminated grain bins in farms, warehouses, and processing plants throughout the world.

Shed Human Skin—The human body sheds a layer of dead skin cells every month, and mold grows on those cells—on clothes, bedding, and upholstery. Washing clothes and fabrics often, using a detergent that contains borate can curtail mold growth, preventing it from ruining fabrics.

Sports Drink and Water Bottles—These plastic bottles are sometimes re-used repeatedly. After every use, they should be washed, rinsed, and thoroughly dried out. If straws are used, a new one must be used each time. Sports drinks and other sweetened beverages practically invite mold to grow. Drink bottles are often given away as promotional items, but regardless, they are cheap enough to be frequently replaced.

Hummingbird Feeders—All parts of the feeder should be cleaned, especially the inside, before refilling it with sugar water. Otherwise, mold will often be evident within one week on the upper inside surface of the bottle. Bottle-cleaning brushes will usually get the mold off. Rinsing the feeder with hot water will not kill the mold, but it will help loosen it up so that it can be flushed out.

Hummingbirds will drink from feeders with mold growing in them, but they will opt for a cleaner feeder if one is available elsewhere in the neighborhood. When the hummingbirds have left the area for the season, washing and brushing the feeder with a detergent will sufficiently clean it. It can then be rinsed then dried, and stored in a clean dry place.

Coolers—Food and water coolers are wonderful places for mold to grow. Have you ever opened a cooler and found mold growing inside? If the cooler had been thoroughly washed, disinfected, dried, and left open and kept dry during storage, the mold wouldn't have been there.

Houseplants—Indoor plants bring a little outdoor beauty into a home, but molds can thrive on houseplants and in potting soil. Garden supply stores carry chemicals that will kill mold—both on the plants and in the soil—but killing the symbiotic fungi in the soil can have detrimental effects on the plants. Other soil fungi break down the different particulates in the soil into nutrients that all plants require. Unfortunately, there is no "happy medium" here—it comes down to a choice between having indoor houseplants and living with their mold and spores, or not having any indoor plants.

Floors in Bathrooms—The bathroom offers three opportunities for moisture problems; the shower stall/tub, the sink and the toilet. Any water problems in these areas will usually affect the floor, which can spawn a series of events in the structure of the house that could be tragic.

Toilets offer one of the greatest threats for mold growth, and it's not simply that they are sources of continuous moisture. There is a water supply line, which should be maintained, but most mold problems are caused by a leak in the wax ring seal. When properly installed, the wax ring provides a watertight seal between the bottom of the toilet and the flanged drain. If the seal is worn or damaged and sewage wastewater leaks between the floor and the subfloor, then mold can grow unnoticed for weeks, months, or years.

Mold problems in bathroom floors are especially serious—they can cause major health problems for families and pets. One should periodically inspect for any sign of leakage around all fixtures, and, if possible, under the floor. If water damage has occurred, the problem must be fixed promptly while ensuring that the area is completely dried before and after repair.

Air Ducts—In homes with central forced-air heating and cooling systems, the interior walls of the air ducts offer a potential haven for mold growth. Whenever the fan of a mold-infested ventilation system circulates air, more spores spread throughout the house.

There are ways to reduce the spore count in forced-air systems. Newly installed systems are generally the easiest to modify. Products such as in-line electrostatic air filters that can trap airborne particles can often be installed, but mold spores can be as small as one micron wide, and filters capable of removing such miniscule particles are very expensive. It may be difficult to retrofit some systems, especially older ones.

UV light units can also be retrofitted into some systems. If practical, combining an electrostatic air filter and UV light units into an existing air system can drastically improve the IAQ.

For people who are especially susceptible to airborne particulates, forced-air systems are not the best way to heat and cool a house. Electric heat, radiant heat, or hot-water/boiler systems are more appropriate, but each has its own drawbacks. Electric and radiant heat systems are expensive to operate; however, the cost of running these units should not be a reason to deny a family a healthy living environment. Hot water or boiler systems present the risk of leaks in various places in the house, due to the pipe layout.

Paint—Molds feed on the minerals and oils in paint, often causing the paint to discolor. If mold is present, it must not be simply painted over. However, if treated in this cosmetic fashion it may look fine for a while. Eventually, the mold problem will worsen, growing even deeper into the materials behind the paint, causing it to peel off, leaving an ugly and unhealthy mess.

Garden Hoses—When outside, water must never be consumed if obtained from a hose! Besides mold, which is almost always growing in the hose, bacteria and small insects can sometimes get inside it. When hoses are not in use, moisture remains in them for long periods of time. Hoses should only be stored after being thoroughly drained. This will help reduce—but not eliminate—bacteria and mold growth inside the hose.

Leather items—When leather products such as jackets, belts, shoes, and baseball gloves get wet, they become very susceptible to mold infestation. Usually, moldy leather goods should be thrown away. If the mold infestation is minor, a professional dry cleaning may save the item.

Animal and bird dung provide a fertile incubator for mold. This can be a serious health issue for people who raise birds, reptiles, or small mammals. Keeping cages well cleaned is vital.

Christmas trees set up inside the house during the holidays can harbor molds that are already on the tree when it is purchased, or they can start growing after the tree has been brought inside. Allergic or mold-sensitive individuals should consider using clean artificial trees.

"Snow molds" are common on lawns that receive heavy snow accumulations. When the snow is deep, it actually keeps the lawns underneath warmer, promoting mold growth. Raking the lawn after the snow melts will help nature dissipate the mold.

Films, books, audio and **video tapes,** and **vinyl record jackets** are all places molds often grow—especially if stored incorrectly. Store these items in cool, dry places, not in attics or basements.

MOLDS IN RVS, TRAVEL TRAILERS, CAMPERS, AND TENTS

There are more molds in recreational vehicles (RVs), travel trailers, campers and tents than anyone realizes. In an upstate New York campground, an almost-brand-new "fifth-wheel" travel trailer with "pullouts" (sections that pull out from the main trailer to make a room bigger) was found to be severely infested with mold. The sealing gaskets on the pullout failed to keep the water out during the winter and moisture entered on both sides of it. The carpet and the wooden floor were left damp or wet for months. An overwhelmingly musty odor flooded the camper, when the storage area under the dining room seat was removed. The storage area under the seat provided a perfect environment for mold growth. Little black blotches of mold resided all over the inside of the wood frame and the underside of the plywood seat. Mold fully "dusted" the carpet inside the storage area. The plywood seat was replaced, the carpet was removed, the entire area was cleaned with the standard Borax solution (one cup of Borax detergent to one gallon of distilled water) and dried with fans, quickly and completely.

After drying, a mold preventative was sprayed over all of the remaining mold-infested wood. Since the mold had not penetrated the wood very deeply, the wood was cleaned and disinfected, not replaced. Sanding the wood was not a sound idea, due to the risk of spreading the spores within the living space. This area is now being inspected weekly to see if any mold reappears.

Finally, the exterior rubber seal moldings on the pullout were replaced to prevent future leaks.

Skylights on the roofs of RVs, travel trailers and campers are notorious leakers. Dirty skylights can grow mold on them, and they can leak around the seams. Also, when it rains, people simply forget to close their skylights and water gets inside.

From the floor to the roof, a mobile RV flexes somewhat while it is being driven or towed, causing sealed joints to crack and leak.

There are also sewer vents, gray-water vents, air conditioning units, and TV antennas on the roofs of RVs and travel trailers. These are all places where moisture can enter. Inspecting the ceiling materials on the inside for any water stains or soft spots should be performed as a normal routine.

When trailers are not in use, they should have a few side windows left "slightly" open to permit air circulation. Stagnant air breeds mold.

Similar to houses, RVs have toilets, sinks, water lines, drainage traps, and showers—all of which are prime candidates for mold infestation in the event of a leak.

RVs are typically winterized by draining the water lines and pumping an RV winterizing fluid (usually pink in color) in them—in hope of pumping out and diluting the residual water left in the pipes. This protects the pipes from bursting. One problem arises: most of these products contain propylene glycol, a synthetic glycerine that it is like candy to mold!

However, there is an alternative to putting anti-freeze in the water lines. Compressed air forced through the water lines via the water inlet on the outside of the RV will help eliminate residual water. To ensure that a pipe will not break under air pressure, at least one valve or faucet must remain open at all times, to vent the pressure. Water lines should be opened separately: the hot-water holding tank should be drained and blown out first; the hot and cold faucets furthest from the water inlet can then be opened and blown out; then the next faucet;

and so on. Air should be forced into the system until water completely stops coming out of each fixture, including the toilet.

In addition, water in each of the drain traps must be blown out. The use of liquid antifreeze is unnecessary in any of the pipes, as long as all water has been blown out of the water lines and drain traps.

Tents very often have an odor when retrieved from storage, especially after many months in the basement. Tents should be thoroughly dried before they are packed away for the season. A mold-infested tent can be washed in a large tub with Borax detergent and water. Erecting the tent after washing it, allows it to dry thoroughly before it is stored.

BUILDING A HOME
WITH MOLDS IN MIND

If you are considering building a new home and want it to be as mold free as possible, then you should discuss building materials and techniques with your contractor. Mold prevention starts before the foundation is poured and continues well after all the walls are plastered. Here are some ideas to help prevent mold infestations in a new home.

- Consult a mold inspector—or another environmental professional with pertinent qualifications—to supervise the construction, and verify that mold-preventative materials and practices are employed.

- Choose your building location carefully. Do not build in flood-prone areas, near underground springs, or where water settles. Ideal locations for building a home include lots with effective natural underground drainage. Most states require a percolation test prior to acquiring a building permit. You should discuss this with your local building code department.

- Make sure the building lot is sufficiently cleared of all trees, shrubs and brush— moisture can be trapped in dense vegetation.

- Discuss the foundation pouring with your contractor prior to starting the work. Floor drainage is important; a below-the-floor sump pump may be a suitable preventative measure for your situation. The outside walls of the foundation should be waterproofed with a special concrete sealer. Non-porous, closed-cell foam insulation around the outside walls will reduce the heat differential between the house interior and exterior, thus preventing condensation problems. Drainage stone and tiles should be used along the perimeter of the foundation to help keep water at a distance from the building. Have your lot graded properly—a landscaper knows how much pitch is needed to divert water, and has the right equipment to do the job correctly.

- Always inspect the building materials as they are delivered to the building site. You are paying for it, so you have the right to check its quality. Do not use any wood that is wet, damp, or has any water stains. Informing your contractor of your intentions will also be a deterrent against the use of poor-quality building materials.

• After framing, treat all completely dry lumber with a mold preventative made for wood applications. Let it dry, and then spray it a second time. After that dries, spray it again with a long-term preventative fungicide containing metallic oxides. After the fungicide-treated wood has been used in the framing, spray with the preventative fungicide yet again. This whole procedure (which should cost under $4000 for a single-family home under 2500 square feet) will kill all spores and mold that might be on the wood and inhibit the growth of mold in the future. If you choose the right products, other benefits like resistance to termites and ants, and fire will also be achieved.

• Install hardwood floors rather than carpeting. Use throw rugs and clean them outdoors often.

- Use a special concrete sealer that protects from radon and water seepage into basements on interior non-painted floors and foundation walls.

- Use fiberglass insulation instead of the blown-in cellulose type.

- Whenever possible, use modern products such as mold-resistant paneling.

- Insulate all water pipes to minimize condensation.

- If forced-air heating and cooling units are being installed, use only metal for the ductwork. Also, in-line UV lighting units that kill spores as they pass through the system can be installed. These lights will turn on and off automatically. State-of-the-art air filtration systems are also essential in minimizing the indoor spore count.

- On ceilings and walls, use anti-fungal paint, which contains a chemical additive; it can be purchased from any quality paint store.

- Ensure that the attic and basement have adequate air circulation and ventilation.

- Properly install quality double-pane windows throughout the house.

Every year, manufacturers introduce new products to help prevent or to control mold—some have great merit, others do not. Beware of any "secret formulas." A little bit of research will help ensure that your money is not wasted on ineffective products.

AN EXAMPLE OF POOR CONSTRUCTION TECHNIQUES:

This house, under construction, is a prime example of a structure that has not been protected from rain. This house has a high probability of mold infestations—especially if the wood is not dried out prior to putting up the drywall.

PART FOUR
MOLD REMEDIATION
ISSUES

INSPECTION

Houses are the largest investments most people ever make. Isn't it prudent to maintain healthy Indoor Air Quality? Most indoor mold problems result from negligence and/or ignorance on the part of the homeowner; education, common sense, and awareness are the best and least expensive weapons for protecting the substantial investments we make in our homes.

Uncovering mold in a house is really not a difficult task for the homeowner or landlord. However, it is one thing merely to uncover mold and another thing to understand the total scope of the problem and properly remedy the situation. Approach your inspection as if it were a crime scene investigation—take a forensic approach in finding not only the mold, but also the underlying cause of the mold infestation.

If mold is found in a home, do not panic. Assess the situation yourself prior to hiring a certified mold inspector who may be found in the phone book. The local health department may assist in recommending someone with credentials and expertise in this field, or another business that specializes in overall IAQ. Searching the Internet may also help you find a qualified contractor in your area.

Professional mold inspections are not cheap; a certified mold inspector may charge between $250 and $2000 for a professional inspection, including a report of his findings. Protocol plans for additional testing and remediation—and the actual remediation itself—are additional.

The inspection charge varies from contractor to contractor and will reflect the size of the house and the distance the mold inspector must travel. A quality, certified mold inspector has a vast knowledge of the subject matter—an education that the inspector had to pay for. Ask to see his credentials.

A mold inspector should arrive at the scene with all the needed equipment. If he "forgot something" and has to return, be wary—he is probably not a qualified professional. High-tech equipment such as a hidden moisture meter, thermometer, hygrometer, air sampling pump, culture plates, masks, sterile white suits, gloves, and duct tape may be used. The white suits, gloves, and tape (used to seal the clothing) are disposable. The initial cost for this basic equipment can exceed $2000.

Some larger environmental IAQ firms can provide "one-stop shop" services. If they must travel from out of town, the associated costs will be factored into the inspection charges. Typically, an inspection is not a cost estimate for fixing a mold problem, but it will define the extent of an infestation. If the problem is very minor, recommendations for correcting the problem will be verbal. The inspection charge can sometimes be credited toward the generation of a protocol plan; inquire when negotiating the inspection contract. Try to get as much for your money as possible.

Each test that uses a culture plate for collecting and analyzing mold spores can cost $40–$150. The spores that have germinated on the culture plate are allowed to grow into colonies as it is sent to the laboratory for analysis.

Note: The U.S. government does not allow mold or mold spores to be intentionally shipped via the U.S. Postal Service.

Once the existence of mold is confirmed, a person can then assess the degree of infestation and damage; it will then be easier to discuss the problem with a mold inspector or an IAQ professional, if needed.

THE FOLLOWING ARE USEFUL WAYS OF FINDING MOLD IN A HOME AND MAKING DECISIONS ON WHAT TO DO ABOUT IT.

1. **Smell**—When entering a room, sniff the air for mildew, mustiness, or an ammonia-like odor. After the initial detection of an odor, an N95 (minimal protection) breathing mask should be used to protect oneself from inhaling too many spores, especially when moving things around during the inspection. Mold spores are released in great abundance when disturbed. Simply walking past mold provides enough air turbulence to launch spores into the air. Most mold inspectors equip themselves with respirators as soon as they detect a significant odor or if they already know that there is a definite mold problem.

   In heavily infested areas, the odor from a mold's Microbial Volatile Organic Compounds (MVOCs) is quite apparent.

Attics and basements notoriously emit this odor, which is often the first sign of a problem. It can sometimes be difficult for an individual to notice the odor if one is habituated to that environment. Our noses have self-adjusting sensory receptors that will block out common or routine odors after a period of time.

2. **Look**—Slowly and methodically look for any signs of discoloration, swelling, or blistering on the walls, wallpaper, floors and ceilings, and especially in the corners of all rooms, the basement, the cellar and the garage. Since moisture is usually the culprit, water stains are reasonable indications as to what may be behind or underneath those areas. When a stain, water damage, or mold has been found, the source of the problem must be located and quickly repaired or corrected.

A person can assess the damage and infestation to determine the procedure needed for remediation, by following the general guidelines outlined in this chapter. Mold problems involving fewer then ten square feet can often be remedied by the homeowner, with low costs.

During an inspection tour, ensure that all air intake and exhaust vents are operating properly. Dryer vents often build-up lint on the exhausts' trap door, and moist air can be trapped inside the house.

If outdoor water sprinklers are used on a property, make sure they do not spray on the house or along the basement walls. Water can leach down along the outside of the basement walls and find its way through cracks and into the house.

Contact a professional if mold behind a wall is suspected—only he has the equipment to investigate without tearing down the wall. Prior to drilling any holes in the walls to insert a probe, a good mold inspector will determine when the house was built. This is very important, for prior to July 1, 1985, asbestos was allowed in construction materials in the U.S. Asbestos was used extensively in dry wall panels. Canada still allows the use of asbestos in dry wall sheets, so be careful if your materials came from our northern neighbor.

Prior to Jan 1, 1978, lead was allowed to be used in paint. Airborne asbestos and lead paint dust are very hazardous in homes built before these dates—drilling holes in walls should be performed carefully.

The presence of lead in paint can be uncovered by using a kit available in many hardware stores. A typical kit's instructions direct the homeowner to first "rough" up the paint in an inconspicuous area with wet sand paper, and then to let it dry. A gel, included in the kit, is then rubbed onto the roughed-up area. The test area will turn a shade of green in about a minute if lead paint is present.

If there is no asbestos and no lead paint in or on the walls, then half-inch-diameter holes need to be drilled to allow the use of a fiber optic inspection device. This is a flexible 18" scope that is inserted into the wall so that the backside of the wall can be viewed through an eyepiece. More holes may be required if an infestation presents itself or if the area drilled cannot be viewed due to firewalls. Firewalls are cut pieces of lumber that are fitted between studs to help prevent flames from spreading vertically to the upper floor(s).

A professional mold inspector will always fill the drilled holes with spackling compound when finished. This seal prevents any particulates from coming out from behind the walls. The inspector is not usually responsible for sanding or repainting the wall—that is the responsibility of the homeowner (this should be spelled out in the contract for services).

A thorough inspector will numerically label the areas tested and, usually, take photographs for future reference.

Another common technique in a walk-through mold inspection is opening and closing every door in the house. If a door sticks or is out of alignment, another clue has surfaced: water damage can cause settling or warping, which will hinder proper door operation.

A certified mold inspector knows what questions to ask (usually, a checklist is used). He will have the right equipment for the inspection, and he knows how to test for mold spores and how to distinguish between the various types of mold. A certified mold inspector is a public health servant—his job is to help ensure a safe and healthy environment in a home by providing guidance regarding mold problems.

A house is not the only place to inspect for mold. Workplaces and schools are also areas where people spend a lot of time, and countless incidents of serious mold problems occur in these places—especially in schools—that could harm children's health.

EXAMPLES OF MOLD INFESTATIONS

Mold on basement walls—Figures 1 and 2 (next page) show the amount of mold infestation that occurred when water pipes froze and burst during the winter. The water was turned off, but, within six months, mold had infested the entire basement. The estimates to clean up the mold far exceeded the value of the house—so it was demolished! This is a common outcome when a house has been neglected for months after water damage.

Figure 1

Figure 2

This picture exhibits one outcome when a home is flooded. In the U.S., hundreds of homes get flooded each year, and many of them are never properly remediated. Flooded houses can quickly develop heavy mold infestation, and the "Sick Building Syndrome" if they are not properly and completely cleaned up.

Mold on a ceiling tile

Mold Growth Under Kitchen Wallpaper

A MOLD INSPECTOR AT WORK

Calibration of an Air Sample Pump prior to usage

Collecting a Carpet Sampling

A Hidden Moisture Meter used
to detect moisture in the wall.

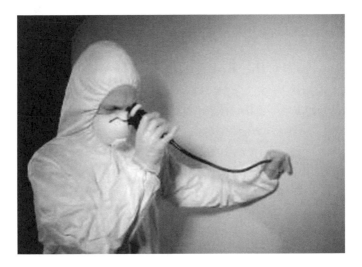

A flexible Fiber Optic viewing device
is used to inspect the inside of a wall.

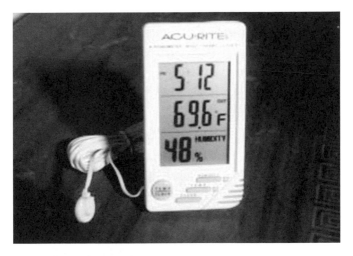

A Relative Humidity Meter and Thermometer.

TESTING AND SAMPLING

Either the homeowner or a professional mold inspector can test for mold. Mold test kits are available in stores and over the Internet. If the homeowner decides to do it himself, he needs to understand what type of information the kit will provide. A qualified mold inspector will know exactly how to accomplish effective testing. He may have to perform multiple tests in the house by collecting airborne spores and mold on surface areas for analysis. Viable and non-viable mold testing methods each have their own inherent limitations—no one test can provide all the answers.

There are presently two types of mold testing methodologies that can be accurately performed: viable (non-volumetric air sampling) and non-viable (volumetric air sampling) testing.

VIABLE TESTING

Viable testing involves the culture (growth) of mold in petri dishes containing a fungal growth medium (substrate). These petri dishes are known as settling, sedimentation, culture, or gravity plates. The petri dishes are opened for a short period of time (often 15 minutes) and inoculated as gravity drops spores from the air onto the petri dish. After inoculation, the spores will usually start growing within 24 to 48 hours (more for some species), and individual mold colonies will begin to appear. The number of colonizing spores or "Colony Forming Units" are then counted, recorded, and mathematically extrapolated to calculate the approximate number of airborne spores in the area/room tested. The mold colonies are each microscopically analyzed by qualified technicians. They usually have strong backgrounds in microbiology, and have been trained to accurately identify the mold colonies to genus and, in some cases, to species. Gravity testing also provides preliminary qualitative information that is very useful in situations such as in the food preparation and packaging industries (for air quality monitoring). The difference between indoor and outdoor airborne spore counts also provides valuable information for making an assessment. Insurance companies and attorneys require outdoor "control" spore counts as baselines. In time, while periodically monitoring (looking through the top of transparent culture plate), individual spores colonizing, will start to appear. Although there is no comprehensive taxonomic database that defines every species of fungus, enough is known about the commonalities within each of the many genera for identification to genus. A good laboratory can provide an accurate analysis in seven to ten days. Results provided by a laboratory in just a couple of days will not be accurate: some species of mold take more than two days to form colonies, so some molds will be omitted.

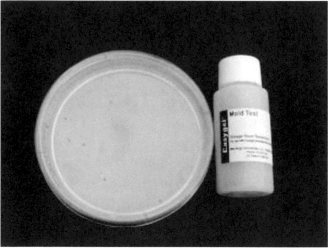

Petri dish and a plastic bottle containing an agar solution

Regardless, gravity testing has other limitations. First, mold growing media such as Malt Extract Agar (MEA) have a shelf life of one to three years. Second, many molds will not grow on some or all types of agar. Third, some mold spores are less frequently airborne due to their size, shape, weight, or stickiness. *Stachybotrys* is a prime example: gravity testing is not an appropriate means to test for *Stachybotrys* mold. Fourth, some of the cultured molds' colonies will not produce spores (sporulate), and a mold cannot be positively identified without studying how it sporulates.

MEA, or simply "agar," as it is commonly called, is usually purchased in a prepared liquid form. After a shallow pool of it has been poured into a petri dish, the liquid agar will gelatinize in about 30 minutes.

Airborne spores fall on the agar in a room, which has been prepared for a 15 (standard) to 60-minute test. The cover plate is then placed on top of the culture plate and a wide piece of tape is used to seal the culture. The inspector labels the bottom plate with his name, client ID, date, sample number, and any other important information. The accompanying data sheet form must correspond exactly to the information on the culture plates—proper protocol and documentation are essential, especially in cases where insurance disputes or legal cases could arise.

Premixed culture plate kits can be purchased through the Internet, a mold inspector or directly from a laboratory.

Another type of viable testing is "swab testing." The area to be tested is gently swiped with a sterile swab that has been dipped in a buffer medium (which comes in a long capsule that has a screw-on cap). The swab is placed in the supplied culture tube, and the cap is sealed. The culture sample is sent within 24 hours to the laboratory, which will continue to grow any mold colonies until they are identified microscopically.

A viable test for identifying specific molds is "surface sampling" or "tape lift sampling." This technique is often employed to identify visible mold colonies, in the home or elsewhere. Using gloves, a piece of clear tape is carefully placed on top of the mold and a sample is lifted off. The tape is then immediately sealed and sent to the laboratory from which the kit was purchased. The charge for this type of testing is often $40, or more. There are various test kits on the market; carefully follow the directions for submitting the sample.

The best way to identify the culprit in visible mold infestations is to perform the tape lift method and immediately prepare a microscope slide for onsite analysis. Unfortunately, very few mold inspectors are skilled at microscopic examination and identification of mold samples, but can only procure the samples.

MICROSCOPIC SLIDE PREPARATION

A SIMPLE METHOD FOR MICROSCOPIC EXAMINATION OF A VIABLE MOLD SAMPLE LIFTED FROM ACTUAL MOLD GROWTH

This temporary slide preparation method works well for a "real-time" look at live mold.

(Note: The mold shown in the following culture plate examples is 10 days old and gloves were not worn for clarity purposes.)

1. Do not touch the mold with your fingers—wear plastic gloves when preparing a slide.

2. Place a drop of water in the center of a standard 1"X 3" microscope slide.

3. Using a one-inch strip of clear tape, lightly press it, glue-side-down, on the mold, and then lift it up. (Do not overload the tape with mold; you don't need much.)

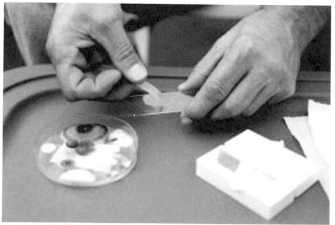

4. Briefly touch the mold side of the tape against the drop of water on the slide.

5. Turn the tape over and place the tape on the slide, mold side up.

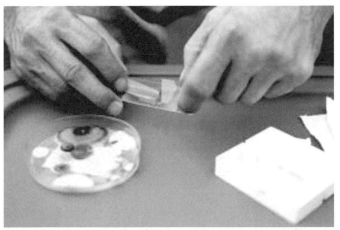

6. Place a standard 22 x 22 mm cover slip on top of the tape/slide.

7. With a tissue or cotton swab, lightly press down on the slip and pat or swab dry all around the slip. Be careful to keep the slip centered on the slide. Finally, seal the slip to the slide by brushing the edges of the slip with nail polish.

8. Place the prepared slide under the microscope and, starting with the lowest-power objective lens, focus in, then move up in magnification to around 400x and refocus.

Although, anyone who has taken a high school biology course should be able to prepare a slide and examine it with a microscope, identifying a mold can be very difficult—even impossible—for an untrained individual. However, it is possible to identify some of the most common types of indoor molds and mold spores with the aid of this book.

NON-VIABLE TESTING

In non-viable testing (volumetric air sampling) spores are collected only to determine whether mold spores are present, and if so, to provide an approximate spore count. This is useful information, but it doesn't identify the mold(s) in the sample.

There are various methods for obtaining an airborne spore count, including the gravity plate method already described and the "grab sampling" method (for air and carpet sampling). The grab sampling method uses a controlled air-testing pump with a calibration unit and intake cassette attachments. The air pump is also called an IAQ air/carpet sampling unit or aerosol-testing pump.

Note: The grab sampling and culture plate testing collection methods can each also facilitate mold identification—not just a spore count.

The air-testing pump "grabs" a sample of the particulates in the air or carpet and deposits them on a special slide, which is then removed from the cassette and analyzed under a microscope. A laboratory technician then counts every spore he sees within a certain area on the slide. The count is then used to mathematically extrapolate a spore count.

Human error can easily interfere with this process. The technician has to be able to differentiate between different types of spores, dust, pollen, and numerous other microscopic particles in the sample, which greatly complicates the process. The technician must tabulate the count, typically using a handheld, thumb-triggered counting device. In short, there is plenty of opportunity for error.

The cheapest "do-it-yourself" test kit costs about $10. One example consists of a small unit that is placed in a room for a period of time as indicated by the directions. The test results merely indicate the presence of mold spores in the air—not much to rely on because mold spores are everywhere! However, high spore counts revealed by a do-it-yourself culture plate test kit offers reason enough to hire a professional.

Some mold inspectors will take two culture plate samples per room and let the homeowner keep one while the other sample is sent to a laboratory. A comparison can then be made between the two samples after the laboratory reports its results. If the petri dish that the homeowner kept has only one or two visible mold colonies and the lab reports ten colonies in their sample, something is wrong!

TEST SAMPLING METHODS

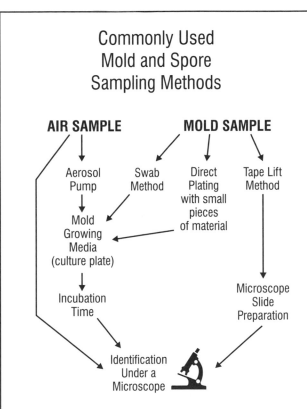

Commonly Used Mold and Spore Sampling Methods

AIR SAMPLE   **MOLD SAMPLE**

Aerosol
Pump

Swab
Method

Direct
Plating
with small
pieces
of material

Tape Lift
Method

Mold
Growing
Media
(culture plate)

Incubation
Time

Microscope
Slide
Preparation

Identification
Under a
Microscope

PROTOCOL PLANS

When mold is found indoors, a plan to eliminate it and prevent it from returning must be established as quickly as possible. If the mold problem is as simple as moldy shower tiles, then the protocol plan is simple. If the mold has infested the entire basement of a home, then the home's value and the estimated cleanup costs must be weighed, and a protocol remediation plan devised based on the objective and the availability or lack of funds.

What should be done if mold is found in a shower stall? In most households, the cleanup usually consists of applying a bleach product or spraying a tub-and-tile cleaner onto the moldy area and wiping it off. The cleanup rags are usually reused.

A better job should be done than that! It is certainly unnecessary to hire a professional remediator to clean a few small mold colonies on the grout, which pose minimal threat to most people. However, a few precautions should be taken. To err on the side of caution, the area should be ventilated (window fans can help) and a basic white filter mask (N95 rating or better), rubber gloves, and an old long-sleeve shirt and long pants should be worn during the cleaning of mold. A borate-based detergent mixed into a liquid paste, or the standard sprayable Borax solution, applied and lightly rubbed with a throwaway sponge, or paper towel, offers the best way to remove minor bathroom mold colonies. The solution should be allowed to dry on the infested areas, and then repeated. It takes about 15 to 30 minutes for borates to kill mold. The cleanup materials should be immediately removed from the house without spreading the spores inside.

The next step is prevention—you do not want mold to colonize in your shower stall again. Clean more often (with a Borax solution) and try to improve the bathroom's ventilation.

For larger, more severe infestations, a protocol plan should be prepared by a mold specialist. This is not usually covered by the inspection fee. A thorough plan requires an engineering approach to the problems—a step-by-step "unlayering" of the infestation must be intelligently formulated.

A typical protocol plan for a large mold infestation can run from $500 to $2000 or more. This does not include the cost of remediation or follow-up clearance testing. Any contract for the development of a protocol plan (as for inspection or remediation) should be in writing, and signed by both parties.

If a mold problem exists, think about what should be done before acting. Do not just start scrubbing mold; during the "quick and easy" cleanup, you may inhale thousands of spores!

If you have a mold problem larger than 10 square feet, you need professional guidance.

Remediation

Once a protocol plan has been formulated, remediation should begin as soon as possible to get rid of the mold and provide a healthier indoor living environment.

Various organizations use different descriptions of levels of mold contamination. For example, the New York City Department of Health defines five different "abatement" levels according to the physical size (in square feet) and type of infestation. **A Small Isolated Area** is the only category for which an untrained individual should remedy the problem. Use of Personal Protection Equipment (PPE) such as goggles, a breathing filter mask, rubber gloves, and full body suits (old clothes will do) are a minimum for cleanup.

The other categories address worse infestations that should always trigger inspection, testing, remediation, and "clearance" testing (follow-up) by trained professionals using the right equipment; otherwise, there is great potential for making matters worse by spreading spores throughout a larger area.

People who are susceptible to allergies should not clean up mold infestations.

173

REMEDIATION CATEGORIES

LEVEL I

Small Isolated Areas—Less than 10 square feet of surface infestation is generally considered to be a small contaminated area. (A typical house door is about three feet wide and seven feet tall—21 square feet.) Shower stall infestations usually fall within this category. Shower stalls will readily grow green or blackish mold in the grouted areas. These types of molds are usually more of an aesthetic annoyance than a health danger, except for people who are especially susceptible or allergic to airborne contaminants. To remedy such common household mold problems, follow the protocol plan outlined previously.

LEVEL II

Mid-Sized Isolated Areas—More than 10 and less than 30 square feet of surface mold is categorized as a medium-sized infestation. A 4' X 8' sheet of drywall is 32 square feet. The protective gloves and mask used for a small infestation are not sufficient for cleaning up infested areas this large. Better masks or breathing apparatus, rubber gloves, full-length hooded suits, and protective footgear should be worn by technicians remediating infestations categorized in Levels II, III, IV or V.

LEVEL III

Large Isolated Areas—Fungal growth of 30–100 square feet of surface mold is considered a large infestation—carpeting, shower stalls, or wall panels in this class often need to be professionally removed after the mold has been killed.

LEVEL IV

Extensive Contamination—Infestations of more than 100 contiguous square feet are very serious. In cases of this magnitude, entire walls, floors and ceilings need to be professionally excised and replaced. In some cases, the associated buildings need to be demolished.

LEVEL V

Remediation of HVAC (Heating Ventilation and Air Conditioning) Systems—Mold contaminations involving an area greater than 100 square feet fall into this category. If carpeting is mold-infested, then forced-air heating and cooling systems also need to be cleaned or even replaced. In most cases however, technicians with the right equipment can properly clean the air ducts and coils in HVAC systems.

FOUR STEPS IN REMEDIATION

- **Test** for mold and mold spores
- **Kill** the mold
- **Remove** the mold
- **Prevent** mold from returning

A mold remediator should be certified, licensed, insured, and bonded. He should be in full compliance with the American Conference of Governmental Industrial Hygienists for mold remediation techniques.

REMEDIATION GUIDELINES

*All professional mold remediators
should follow the proceeding guidelines:*

1. A contract, approved and signed by both parties, must include authorization for chemical usage. (The mold inspector may monitor the remediation to ensure that the protocol plan is rigorously followed.)

2. All federal, state, and local codes and statutes must be understood and followed.

3. All molds must be killed with a fungicide/
    disinfectant before removal. This could
    mean "tenting" the house or creating a
    "positive pressure environment" for a
    period of time, to kill spores in the build-
    ing.

4. Technicians must wear full-body Personal
    Protective Equipment (PPE) during the
    remediation process.

5. All furniture and other items must be con-
    tained and then removed from the work
    area prior to mold remediation. They
    should all be cleaned separately or dis-
    posed of, in accordance with the protocol
    plan.

6. If necessary, as specified in the protocol
    plan, the remediator must contain, or
    quarantine the infected area with a
    double layer of 6 MIL (.006") plastic
    sheeting and duct tape, creating an air-
    tight work area. The most direct route
    should be used for removing bagged ma-
    terials from the building. Long zippers,
    duct-taped to each layer of plastic sheet-
    ing, should be used by the technicians to
    get in and out of the work area.

7. Industrial-strength HEPA filters and dehumidifiers should be used during remediation. A negative-pressure work area employing a "push-pull" fan/filter method should be used. One unit pulls air in from the outside and another (located at the opposite side of the room, if possible) pushes the air out of the room. The "push" (exhaust) unit is set higher than the "pull" (intake) unit. This creates a negative-pressure working area while filtering the outgoing air to prevent spores from being spread outside the work area.

8. The double plastic bag method should be used for disposal of infested materials. Large pieces of material should be carefully cut or broken into smaller, more manageable pieces. Infested materials should be sealed in large zippered plastic bags, and each then placed into another bag. All infested materials must be properly decontaminated or disposed of, according to the protocol plan.

9. After all the mold has been removed, a disinfectant mold killer should be sprayed on all remaining areas, allowed to dry, vacuumed with a HEPA unit, and the entire process then be repeated a second time.

10. After the second coating of disinfectant has been dried and HEPA vacuumed, an antimicrobial coating should be sprayed on all areas, allowed to dry, and then repeated a second time, to prevent future mold growth.

11. For medium and large mold infestations, clearance testing should be performed before the homeowner is permitted to return. Once doors and windows are opened and people are allowed to enter and exit the dwelling, mold spores will re-enter; testing performed after that will be of little value in assessing the remediation efforts.

12. After clearance testing has been successfully completed, the homeowner should be advised regarding how to keep the environment in the house as mold-free as practical. This can include "subscription" testing, which could consist of a two-year plan for follow-up testing of airborne mold spores.

13. Carpets and ductwork must be properly cleaned after the removal of the mold if called for in the protocol plan.

EQUIPMENT

Many different kinds of cleaning solutions and equipment are used in cleaning up mold. Most of which have already been discussed and used as examples, but there are some more specifics.

There are two different kinds of mold disinfectants: Sporicides and Fungicides. **Sporicides** kill spores. **Fungicides** kill live molds, and most will also kill mold spores. Consult with a retailer or an environmental contractor—and always read the labels.

New products for cleaning and removing mold hit the market almost every year—keep an eye out for them. "Environmentally Friendly" solutions for indoor use such as products using "borate" in them, as the major ingredient, are in high demand.

Pressure washers can be used to remove mold from wood such as patio decks. After thoroughly drying the wood, a fungicide should be applied, and allowed to dry before applying a final surface coating that protects the deck from absorbing water. If not performed correctly, the infestation will start all over again.

Proper equipment is vital for inspecting, testing, and removing mold. However, they must be used correctly. Some test products must be positioned correctly in a room. For example, the standard for aerosol pump spore sampling requires placing the cassette three feet above the floor before sampling—this is the area from which we breathe the most air.

Homeowners need not buy expensive equipment that will be used only once or twice. Hand-held hidden moisture meters are high-tech tools that can penetrate three quarters of an inch into cellulose materials and display moisture levels, but they cost about $500. A qualified mold specialist knows what equipment is required to do a job, and should have all the necessary tools.

One of the largest polluters of indoor air is the vacuum cleaner. Most vacuum cleaners pick up the dirt you can see, but the minute particles—including mold spores—are not filtered out; they are simply spewed back into the air.

A high-tech laser particle counter measures exhausted particulates on a particle-per-cubic-foot (ppcf) basis. Unbelievably, many HEPA-type vacuum cleaners spew out between 20,000 and 1,500,000 ppcf, but the best vacuum cleaners release less than 1,000 ppcf. These units are expensive (about $500), but they are excellent; many hospitals use them!

Look for units that advertise the following: "99.97% Filter, removes 0.3 micron size particles," "ULPA Filter," "Allergy Type" or "Healthy." Be sure to read the fine print! Keep the filters clean (or replace them) and empty out the bags frequently. Do not purchase a unit that does not have a sealed collection bag inside.

Central vacuum systems are one of the best ways to reverse the build-up of spores inside of the house. When properly installed, these units vent the exhaust outdoors, away from any air intake units, windows or doors.

PREVENTION

Preventing mold from invading your home, apartment, or RV can be a valuable step toward preventing unwanted health problems—and it may even increase your property value if you can show that your house is "mold-free," based on testing and certification. In New York State, a property disclosure law went into effect in 2002, which requires the seller to declare, in writing, any problems with the home— past or present. Similar laws are being constructed throughout the United States.

All people have been exposed to many types of mold spores and infestations in their lifetimes. However, prevention—including getting tested by a physician for mold allergies— is the key.

HERE ARE SOME MORE STEPS YOU CAN TAKE TO PREVENT MOLD FROM INFESTING YOUR HOME AND DISRUPTING YOUR LIFE.

1. Eliminate any moisture or water problems immediately.

2. Use a dehumidifier in the basement to keep it dry.

3. Keep rooms well ventilated to minimize moisture and reduce stagnant air.

4. Store items properly—allow some space between boxes and other storage containers. Do not place cardboard boxes or other cellulose-based items directly on concrete floors.

5. Thoroughly clean or discard any water-damaged items.

6. Avoid direct body contact with existing mold.

7. Use hygrometers to monitor a room's relative humidity (RH). Strive to keep the RH at 50% or lower.

8. Ozone generators can kill mold spores in your indoor air, but their use is not recommended. They can actually cause a black substance that can be mistaken for mold to build up on walls. Excessive ozone can also break down certain substances in the home, such as the glue used in the manufacture of plywood and particleboard.

9. Use mold-resistant paneling in wall construction.

10. Use HEPA vacuum cleaners for daily cleanup.

11. Clean and disinfect the inside of the refrigerator regularly, and immediately, after any spills.

12. Check fruits, vegetables, and other food for any sign of mold and immediately remove any infested items from the home.

13. Inspect concrete walls and floors to ensure that there are no cracks.

14. Do not install carpeting on concrete floors.

15. Empty the dehumidifier's water collection reservoirs frequently or provide continuous drainage.

16. If you own an RV, camper, or travel trailer, regularly inspect for any signs of leaks, especially if you detect a musty or foul ammonia-like odor inside closets or storage areas. Most leaks originate from roof damage, cracked seams on the edges, and worn-out seals on doors, windows, vents, and pullout sections.

17. Fix any leaks immediately. Check sinks, toilets, tubs and showers frequently. Check pipes to ensure that they do not sweat. Watch for any signs of leaks in roofs, walls, and floors.

18. Research products that claim to eliminate mold spores. The use of simple ionizers, for example, will neither kill mold spores nor eliminate them from the air.

19. Close windows whenever it rains—some mold spores become airborne during rain—but in dry weather, open windows whenever possible.

20. Tightly close containers of foods such as cereals, crackers, snacks, bread, and flour before storing them. Properly put away food soon after eating. Keep beverages capped when you are not drinking them.

21. Use HEPA room air filters; clean or change the filters often.

PART FIVE
WRAP UP AND
RESOURCES

WRAP UP AND RESOURCES

Thank you for purchasing *Black Mold Your Health and Your Home.* If you have mold, I hope this book will help you to more effectively handle your situation—and more importantly—will help you and your family to avoid any potential health threat from mold.

I have gathered the information within these pages from a great many sources. The following especially helpful references are for readers who are interested in learning more about molds.

Thanks Again,
Richard F. Progovitz

MOLD RESOURCES

All Real Estate Trades—Edward Hemway, Mold Instructor and mentor of the author.
Web site: http://www.allretrades.com/

Certified Mold Inspectors and Contractors Institute (CMICI)—"Mold Buster Tips" - Phillip Fry, 2002 from their Mold Certification Program and from their website:
Web site: http://www.certifiedmoldinspectors.com

New York City Department of Health and Mental Hygiene
Web site: http://www.ci.nyc.ny.us/html/doh/

U.S. Environmental Protection Agency
Web site: http://www.epa.gov/iaq/pubs/

U. S. Federal Emergency Management Agency (FEMA)
Web site: http://www.fema.gov/

University of Iowa College of Public Health, Great Plains Center for Agricultural Health
Web site: http://www.public-health.uiowa.edu/GPCAH/

Delaware Health and Social Services, Division of Public Health
Web site: http://www.state.de.us/dhss/dph

North Carolina State University, college of Agriculture and Life Sciences
Web site: http://www.ces.ncsu.edu/depts/

North West Fungus Group
Web site: http://www.fungus.org.uk/nwfg.htm

Centers for Disease Control and Prevention
Web site: http://www.cdc.gov/

Colorado State University, Cooperative Extension
Web site: http://www.ext.colostate.edu/

Fungi Perfecti
Web site: http://www.fungi.com/index.html

U.S. National Library of Medicine
Web site: http://www.nlm.nih.gov/

The American Conference of Governmental
Industrial Hygienists, Web site: http://
www.acgih.org/resources/links.htm

Mold Remediation: Standard Operating Proce-
dures Manual
Web site: http://www.mold-abatement-sop.com

University of Nebraska, Institute of Agriculture
and Natural Resources Web site: http://
www.ianr.unl.edu/PUBS/

U.S. Department of Agriculture
Web site: http://wheat.pw.usda.gov/ggpages/

U.S. Wheat and Barley Scab Initiative
Web site: http://www.scabusa.org/

Purdue University, Botany and Plant Pathology
Web site: http://www.btny.purdue.edu/Extension/

Michigan State University
Web site: http://www.msu.edu/user/

HM Government, Department of Environment
Food and Rural Affairs Web site: http://
www.defra.gov.uk/

Enviro Health Environmental Home Inspections,
Inc.
Web site: http://www.create-your-healthy-
home.com

American College of Occupational and Environmental Medicine
Web site: http://www.acoem.org/guidelines/

New Zealand Dermatological Society
Web site: http://www.dermnetnz.org/index.html

International Code of Botanical Nomenclature
Web site: http://www.bgbm.fu-berlin.de/iapt/
nomenclature/code/tokyo-e/

National Allergy
Web site: http://www.nationalallergy.com

Asthma and Allergy Foundation of America
Web site: http://www.aafa.org

ABOUT THE AUTHOR

Richard F. Progovitz, the grandson of Polish immigrants on both sides, was taught his first mycology lessons in the woods—picking mushrooms.  As a retired mechanical engineer, he is the holder of two patents, a dozen inventions, and the author of ten published technical papers. He has served as president and newsletter editor of the Susquehanna Valley Mycological Society and vice president of the Northeast Mycological Federation. Mycologists also know him as the creator of two sets (Edible and Poisonous) of "Wild Mushroom Identification Flash Cards." As a certified mold inspector and contractor, he frequently lectures on various topics. These informative classes include training people to identify mushrooms, molds, and other fungi, for a variety of organizations

Mr. Progovitz lives with his wife and three children in Johnson City, New York.

You can contact Richard F. Progovitz at:
www.fungibyprogy.com

INDEX